BE A WIZARD
WITH NUMBERS

101 WAYS TO COUNT YOURSELF SMART

"Numbers rule the Universe."

Pythagoras (*c.*570–*c.*490BC)

BE A WIZARD WITH NUMBERS

101 WAYS TO COUNT YOURSELF SMART

ANDREW JEFFREY

DUNCAN BAIRD PUBLSHERS

LONDON

BE A WIZARD WITH NUMBERS
Andrew Jeffrey

Author's dedication: To Alison, William and Daniel: the people who really count.

First published in the United Kingdom and Ireland in 2009 by
Duncan Baird Publishers Ltd
Sixth Floor, Castle House
75–76 Wells Street
London W1T 3QH

Conceived, created and designed by Duncan Baird Publishers

Managing Editor: Caroline Ball
Editor: Katie John
Managing Designer: Clare Thorpe

British Library Cataloguing-in-Publication Data:
A CIP record for this book is available from the British Library

ISBN: 978-184483-832-5

10 9 8 7 6 5 4 3 2

Typeset in Trade Gothic
Colour reproduction by Scanhouse, Malaysia
Printed and bound in China

CONTENTS

INTRODUCTION

Mathematics is an art, a science and a language – properties shared only, perhaps, with music. These qualities make the world of numbers a fascinating and attractive place, yet many otherwise intelligent people have an unnecessary and wholly avoidable fear that prevents them from discovering its power, its beauty, its patterns and its joy.

CAN'T DO … OR CAN'T DO YET?

When I was asked to write this book, I had no hesitation in agreeing. Its central premise seemed to be so much in tune with everything I've come to believe about the ways in which we can view and learn about mathematics. So often, people assume that it's too late to learn to love numbers – if you struggled with the subject at school you may believe, quite wrongly, that you'll never "get it". You may even take comfort in believing that you can't deal with numbers because of an inborn "number blindness". (This does exist – see page 130 – but is far from the most likely cause of your difficulties.) I've lost count (ha!) of the number of people who tell me that, for them, mathematics was a chore rather than a joy, usually because of the style in which it was taught. And, naturally, if we don't grasp the basics of something, we will struggle with what follows. But it's never too late to start, and I felt that this book would provide an excellent opportunity to redress the balance and attempt to show how much fun numbers can be.

The aim of the book is to help you to develop enjoyment of and confidence with numbers, often in surprisingly simple ways. For example, can you count to 1,000? Have you ever done so? Assuming that your answer to the first question is Yes and to the second is No, the next question must be: What makes you so sure that you could definitely do something that you've never actually done? The answer is that, even without realizing it, you're recognizing that numbers form patterns. How these patterns fit together and relate to each other is at the heart of mathematics. What was once a tricky pattern for you to grasp is now so simple that you don't even think about it. Therefore, you should never be afraid of the mathematics you cannot yet do.

THE MAGIC OF NUMBERS

You'll find references to patterns throughout the book. The ability to spot a pattern and then use it to work out what comes next is a crucial skill, whether the pattern be numbers, letters, shapes or colours. As a simple demonstration, consider this pattern:

Red, Black, Red, Black, Red, Black, Red, Black, Red, Black ...

What is the next colour?
What is the 100th colour?

The 100th colour is Black, of course, but how do you *know*? Because 100 is an even number, and every second colour is Black. By reasoning along these lines, you're already thinking mathematically. Although this may be a trivial example, it hides a greater truth: that having discerned the pattern, we can state with confidence what lies further along the sequence – that, for example, the 1,286,295th colour in the sequence would be Red (odd number). It's this ability to make predictions with absolute certainty that gives mathematics much of its power and beauty. As Einstein said: "Pure mathematics is, in its way, the poetry of logical ideas."

As a professional magician, and latterly known in schools around Europe as The Mathemagician, I've always enjoyed mathematical tricks, and during my 20-year teaching career I would frequently use magic tricks to explore new concepts with my school classes. I was surprised to see how popular these lessons seemed to be, and how the tricks helped many children to engage with difficult concepts in a manageable way. They could undoubtedly do the same for you. To help you to become a real numbers wizard, I've included here a few simple and surprising mathematical magic tricks you can try out on your friends and family.

THE REAL, THE RELEVANT AND THE SIMPLY REMARKABLE
The tricks, the puzzles and the exercises are all designed to test and further your understanding. Hopefully you'll start to find, as you work through them, that a lot of mathematical ideas you previously considered too difficult are nothing more than extensions of far simpler ideas that you *do* understand.

For example, you'll learn that if you can double and halve, and multiply and divide by 10, pretty much any calculation that you're likely to encounter on a day-to-day basis can be done in your head. As you'll then discover, this can be of great benefit when shopping, banking, or doing any number of everyday tasks that require the use of numbers – you'll be able to calculate a waiter's tip without embarrassingly having to resort to a calculator, and you'll know whether a deal is as good as it sounds.

But as well as learning plenty of shortcuts with which to make life a little easier, you'll meet some amazing numerical facts, from how people coped before the invention of zero, to a snowflake with an infinite perimeter, to when not to believe your calculator.

The common classroom cry of "But I'll *never* need this when I grow up" is often the literal truth, but once you discover the fascination of numbers you're likely to become intrigued by the far-reaching influence they have in all areas of life. Numbers are everywhere. Politicians and marketing experts often seek to convince us, or justify their decisions, on the basis of statistics. Chances and odds are not always what we expect. By the time you reach Chapter 5, you'll be ready for an insight into how numbers really do affect the way the world works.

I hope that this book will not only appeal to everyone who wants to overcome a sense of numerical inadequacy, but also fire the enthusiasm of those of you who are already comfortable with numbers. Wander along some of the weirder avenues down which numbers can lead you, and enjoy the fun of their magic properties. I've certainly rediscovered long-forgotten facts and tips, gained new knowledge and learnt that some things I believed were, in fact, not entirely correct. I hope that this will mirror your own experience.

So – expect to be challenged, surprised and fascinated as you discover (or rediscover) the wonderful world of numbers.

CHAPTER 1

NUMBERS: FEARSOME OR FUN?

If you're someone for whom maths lessons were the low point of the school week, or who groans when faced with working out how much to tip or what a change in interest rates means, you may have a quite unwarranted fear of numbers. However, numbers are actually very reassuring things to work with because they are so predictable and stable. For example, 64×15 comes out at 960 every time: it doesn't shift according to mood or interpretation. This chapter looks at how you think about numbers, and introduces you to some of the ways in which you can make them work for you.

THE TROUBLE WITH NUMBERS

So often we assume that it's too late to learn to love numbers — if you struggled with maths at school you may believe, quite wrongly, that you'll never "get it". But it's never too late to start to see the underlying patterns in numbers and how they fit together, and to use them accordingly. This book is aimed at helping you to achieve that goal.

SEEING NUMBERS FOR WHAT THEY ARE

Suppose that 114 people want to enter the men's singles competition this year at Wimbledon. There will have to be some qualifying matches, before getting down to the last 64, 32, 16, 8, 4 and then 2 for the final. If you were asked to work out how many matches would be needed, how would you go about it?

Probably, you'd work out how many matches have to be played in the first round, then work out how many matches must be played in each subsequent round and add all the totals together. Mind-boggling, and open to error. But there's an easier way. First ask yourself: what's the final goal? To have one player win the championship. So that means 113 players need to be eliminated. And since one player is eliminated per match, there must be exactly 113 matches played. Simple!

To assume "I can't" is a natural human tendency with anything that we fear, and it's especially true of mathematics. We've all been taught in very different ways, and have doubtless developed differing concepts about numbers. Consequently, we often carry around ill-founded beliefs about them and about our presumed lack of ability. Misunderstandings like this make numbers, and mathematics, appear more difficult than they are. Even highly intelligent people often have blanks when it comes to numbers, and yet with a bit of understanding of the *patterns* of number (rather than some half-remembered, meaningless rules from school) you can find yourself much more likely to succeed. Maybe instead of "I can't" we should start thinking "I can't yet, but if I can work out a pattern of what's going on here ..."

HOW AT HOME ARE YOU WITH NUMBERS?

The following quiz is unusual in that the answers are not the most important thing. What really matters is how you approach the question: do you reach for a calculator, or do you spot a shortcut? For each problem, make a note of how you solved it. Once you have done that, an analysis of your methods will reveal just how comfortable you are with numbers. Good luck, and remember, this is not a test in which you need to score full marks – you may surprise yourself!

1   What is 23 multiplied by 99?
2   $300 \div \frac{1}{3} = ?$
3   How many 1cm cubes would it take to build a larger cube, whose sides were all 7cm?
4   A farmer packs eggs in boxes of 48. How many boxes will he require to pack 472 eggs?
5   What is the total of all the numbers from 1 to 20?

The answers and, most importantly, why:

1   **2277**

If you performed the multiplication correctly, award yourself **3 points**. If you tried but didn't reach the correct answer, award yourself **2 points**. If you spotted that this is very nearly 23×100, so the answer would be $2300 - 23$, award yourself **4 points**.

2   **900**

If you answered 100, you're not alone: division is a much misunderstood concept. It's perhaps most easily thought of as "share into groups of": imagine 300 pizzas, each divided into thirds, and it's easy to see that there are 900 slices. Award yourself **2 points** for attempting any answer at all, **3 points** if you used a calculator, and **4 points** for getting the correct answer by any means other than a calculator.

3   **343**

If you answered by imagining a square 7×7cm layer made up of 49 small cubes, and realized that you'd need seven such layers to make the large cube, have **3 points**. You are a visual learner. If you did it by some other method that involved calculating $7 \times 7 \times 7$, also score **3 points**. If you used a calculator and got it right, score **2 points**. If you tried but got a wrong answer, score **1 point**. The large cube illustrates the way mathematics works — we can start with something very simple and build on our knowledge.

4 **10 boxes**

If you got it right by working out $472 \div 48$, either on paper or using a calculator, award yourself **2 points**. If you got a wrong answer, award yourself **1 mark**, unless your answer was 9.83333. In this case, award yourself **no points**: common sense says the answer must be a whole number. This demonstrates a problem with the way a lot of mathematics is learnt: we forget to ask ourselves whether the answers we come up with actually make sense, often owing to our own lack of confidence with numbers. If you reasoned that 480 eggs would need 10 boxes, but that 9 boxes would not be enough, award yourself **3 points**

5 **210**

If you got this right by adding up the numbers in ascending or descending order, award yourself **2 points**. If you got it wrong using this method, award yourself **1 point**. If you added the numbers up in a different order to make things a bit easier for yourself, give yourself **4 points**, even if your answer wasn't correct. (See page 24 for how Carl Friedrich Gauss confounded his teacher by adding up the numbers from 1 to 100 in under a minute!)

Add up your scores.

If you scored between 0 and 6 points, you probably didn't try to answer every question. This could be for various reasons, including lack of confidence, impatience to reach the end or getting stuck.

The reason why points have been awarded for wrong answers is that it's important for you to realize that wrong answers are simply mistakes: a wrong answer is infinitely better than no answer at all. Often, we don't fail because we tried – we fail because we didn't dare to try.

If you scored between 7 and 12 points, you may well have a few methods that you remember from school and like to rely on, but perhaps were never taught exactly why those methods work, or whether there might be simpler or more appropriate ways to solve questions of these types. This book will certainly help you tackle such problems. Your score also suggests that you were not afraid to tackle questions that you weren't sure about – a very good sign.

If you scored between 13 and 20 points, you're probably a lateral thinker, and will certainly enjoy many of the mathematical ideas in subsequent chapters.

The unusual scoring system highlights how important it is to *have a go*, and also that the way you approach numbers makes a big difference to your chance of success. Spotting shortcuts, recognizing what's actually being asked, not being overwhelmed by apparently "difficult" numbers – these skills all help to demystify maths and will help to make you a number wizard.

Note: there's nothing wrong with calculators: I use mine all the time, but the skill is in being selective … as we shall see in Chapter 2.

NUMBERS AS A CONCEPT

One of the biggest breakthroughs for people who have previously struggled with numbers is the understanding that the human brain does not actually think in numbers at all, but entirely in pictures.

NUMBERS AS INDEPENDENT ENTITIES

In order to understand something fully, the brain needs to build up a mental image of that concept. If we struggle to manipulate numbers, it's usually because we can't really envisage what numbers "look like". We need to build what's known as a "concept image" of each number in our heads.

When very young, we learn that, for example, "3" describes something about a set of objects (its quantity). But this knowledge only gets us part of the way, by enabling us to use numbers as adjectives: 3 birds, 2 eyes.

The next development, which can take place at any age but typically occurs between the ages of four and seven, is to move from thinking of numbers as merely describing a group of things to understanding them as independent concepts. It's hard even to add 23 and 13 if we can only think of these as collections of objects, but once our brain can visualize numbers as ideas *in their own right*, working with numbers becomes much simpler. If we haven't made this transition in thinking, processing numbers becomes meaningless and hence very difficult.

REPRESENTATIONS AND RELATIONSHIPS

Just as our relationships with people become deeper the more we get to know them, so our understanding of numbers improves as we spend more time experiencing them in a variety of different ways. If we limit our experience of 5 to only "the number of ducks in a song", it will be hard later to see that 5 can also be, for example, half of 10 or 1 per cent of 500. This is good news for those who have previously struggled with numbers: expanding your experience of the various facets of numbers will increase your confidence in working with them.

5 OR V OR 五

The shapes that we use for numbers (1, 2, 3, 4, 5, etc) have no intrinsic meaning – only the meaning that we attach to them. The same meaning, or value, may be represented in different ways – the number 5, for example, by spots on a die or the digits on an outstretched hand. In the heading above, we have the modern Western, ancient Roman and Chinese equivalents. Conversely, the same number may mean different things: a monetary value, a time of day, a team number, and so on.

Think about the following: *(solutions on page 132)*

- If 3's opposite is 4 and 6's opposite is 1, what is 5's opposite?
- Which of the following is the odd one out:
  60 ÷ 12
  5,550.55 − 500.55
  40% of 12.5
  10,002 − 9,997
  1hr 44min 28sec plus 3hr 15min 32sec?
- When is it true that 5 = 41? (Or, as a clue, when is it -15?)

HOW NUMBERS BEHAVE

Familiarizing yourself with a city by spending a lot of time there increases your chance of finding efficient shortcuts. You begin to build up a map of the street patterns in your head, and can see links that you never knew about before. The same is true of numbers.

THE PREDICTABILITY OF NUMBERS

Perhaps one of the most crucial keys to understanding numbers is the appreciation that numbers are arranged in patterns, and follow very simple rules indeed. Once these patterns and rules have been grasped, they are never forgotten.

The patterns that exist within numbers are legion – they range from the very simple to the extraordinarily and beautifully complex. We will explore some of these patterns here and throughout this book, and mastering them will help you far more than struggling to remember what you were probably taught at school.

The following patterns of calculation make it simple to break down even the largest numbers – some may already be familiar to you.

- To multiply whole numbers by 10, simply add a nought, and to multiply by 100, add two 0s.
- If the last digit of a number is divisible by 2, then the whole number, however long, is even and therefore divisible by 2 without leftover fractions.
- If a whole number ends in 5 or 0, it's divisible by 5.

- Two odd numbers always add up to an even number.
- If a whole number's digits add up to a number divisible by 3, the entire number also divides by 3 (so you can tell at a glance, for example, that 287,511 is exactly divisible by 3).
- Similarly, if the digits of a number add up to 9 or a multiple of 9, that number will be divisible by 9.
- Halve the last 3 digits, then halve again. If the answer is even, then the original number will divide by 8.
- Remove the last digit, double it, then subtract it from what's left of the original number. Repeat this step until only one digit remains. If this is -7, 0 or 7, the original number is divisible by 7.

Telephone trick

Even your phone number has predictable properties. Write down the last six digits. Rearrange them to form a different six-digit number, then subtract the smaller number from the larger one. Now add up the digits of your answer – do you get a total of 27? This happens an amazingly high number of times; other rarer possibilities are 18 or 36, or very occasionally 9. But the answer is always a multiple of 9.

Here is an example based on the number 731117:

Rearrange the digits to form 317171.

$$731117$$
$$-317171$$
$$413496$$
$$4 + 1 + 3 + 4 + 9 + 6 = 27$$

CHOOSE A NUMBER

With a little knowledge about how numbers behave, you can use your calculator in simple predictive tricks. Try the following with a friend.

Mind-reading magic

Without your friend seeing, write down the number 37 on a piece of paper and turn it face down. Ask them to take the following steps.

- Key a single digit into a calculator three times: eg 555.
- Add up the three digits (eg $5 + 5 + 5 = 15$).
- Divide the three-digit number by the new number (eg $555 \div 15$).

Show them what you wrote down.

The clue is that numbers with three identical digits are all multiples of 111, and $111 = 37 \times 3$. When you add the digits together, it's the same as multiplying your starting digit by 3. When you carry out the last step, you effectively divide the three-digit number by your original digit, taking you to 111, and then divide again by 3, which gives 37.

The power of 9

- Choose a single digit. Multiply it by 9 in your head.
- Use your calculator to multiply your answer by 12,345,679.

The calculator will show your chosen digit repeated nine times.

This works because of a simple but little-known fact: that $12,345,679 \times 9 = 111,111,111$. All you've done is to multiply $111,111,111$ by your chosen digit ... so, for example, starting with 8 will give 888,888,888.

INSTANT GENIUS

It is the year 1784. As the teacher hands out the day's mathematics assignments, he notices with irritation that seven-year-old Carl Friedrich has already finished. He decides to set him something so difficult that it will surely keep him quiet for the whole day.

*   *   *   *   *

"Now then, Gauss, before you go home today, I want you to add up all the numbers from 1 to 100. When you have finished that, you can go –"
"Five thousand and fifty, sir."
"*What*?"
"Five thousand and fifty, sir. Isn't that what you made it?"
"But, I, well, I, um – never mind, off you go, Gauss."

*   *   *

So how did Gauss achieve this, centuries before electronic calculators?
- Imagine the numbers 1 to 100 written along a piece of tape, with 1 on the left and 100 on the right.
- Now imagine a second tape, laid below the original, with the numbers going the opposite way, so that 1 lies below 100, 2 lies below 99 and so on. This gives you 100 pairs of numbers. And – the vital point – each of these pairs adds up to the same number: 101.
- So the total of all these pairs of numbers = $100 \times 101 = 10,100$.
- Since we only want to know the total on one tape, we simply halve this to find our answer: 5,050.

In fact, Gauss only used one line of numbers, which he mentally folded in half. This is similar, but harder to visualize. C.F. Gauss (1777–1855) grew up to become one of the greatest mathematicians of his generation.

NUMBER PATTERNS

Pattern pervades mathematics. The times tables, number sequences and progressions, Latin squares and other predictable patterns provide the basis for everyday number-related things as varied as Sudoku, barcodes and map scales. Understanding the rules behind the patterns lies at the heart of all mathematical learning.

MAKING USE OF PATTERNS

The simplest of all patterns is the counting sequence 1, 2, 3, 4, 5, 6, etc. It is our understanding of pattern that gives us the confidence to be sure that, if we could be bothered, we could count to 1,000 or 1,000,000 without much difficulty. Patterns can also aid calculation in sometimes surprising ways, as the following example shows.

The addition balance

Without working them out, which of these sums do you think is the greater:

$$16 + 17 + 18 + 19 + 20$$
$$\text{or}$$
$$21 + 22 + 23 + 24?$$

There are two instinctive responses to this question: either the first, as it has more numbers, or the second, as it has larger numbers. Of course, you may well have guessed that they are both equal. They are an extension of the following pattern:

$$1 + 2 = 3$$
$$4 + 5 + 6 = 7 + 8$$
$$9 + 10 + 11 + 12 = 13 + 14 + 15$$

Can you predict (and check) what the fifth line would be?

LATIN SQUARES

These are squares containing a specific set of numbers or other symbols, in which no symbol must be repeated more than once in each row or column. A simple example is shown below.

1	2	3	4
4	1	2	3
3	4	1	2
2	3	4	1

Sudoku is an example of a puzzle based on Latin squares, and is an indicator of the human brain's fascination with pattern. Further fascinating patterns and number relationships in the book include the Fibonacci sequence (page 48), pi (page 62), the golden ratio (page 64) and barcodes (page 87).

NUMBER SEQUENCES

The most common number sequences are known as "arithmetic progressions", in which successive numbers in the sequence increase or decrease by a constant amount. An example might be:

4   7   10   13   16   19 ...

For clarity, we describe the 4 as "term 1", the 7 as "term 2" and so on. It's sometimes useful to be able to predict what, for example, the 100th term in such a sequence might be. This is achieved by finding a general rule that will describe any term in a particular sequence, including as yet unidentified ones; this generalized term is referred to as the "nth term".

We can see that the difference between successive terms in this sequence is always 3. This is also true of the 3 times table, so it is likely that the sequence is very close to the 3 times table. Comparing the sequence with the 3 times table gives us:

Term number	1	2	3	4	5	6	n
3-times table	3	6	9	12	15	18	3n
Sequence	4	7	10	13	16	19	3n + 1

The last column (n) is an attempt to generalize the pattern. For any number (n), the nth term of the 3 times table will be 3 times n, written as "3n". But the nth term of the sequence is always 1 higher, so we can say that the nth term of the sequence is $3n + 1$. We can now predict with a hundred per cent certainty that the 100th term will be $3 \times 100 + 1$, or 301. What's more, the 200th term will be 601, the 30th term will be 91, and so on. Finding the general rule ($3n + 1$) allows us to find *any* term in the sequence rapidly and easily without having to write it all out.

TAKING THE TERROR OUT OF TIMES TABLES

Often feared and little loved, times tables are grids of numbers representing the sequences with the nth terms "1n" to "10n". Children are encouraged to learn these tables by heart, even before they fully understand their meaning. Those children who spot patterns are far more likely to succeed than those who do not. At first glance, there are 100 tables facts to be learnt:

×	1	2	3	4	5	6	7	8	9	10
1	1	2	3	4	5	6	7	8	9	10
2	2	4	6	8	10	12	14	16	18	20
3	3	6	9	12	15	18	21	24	27	30
4	4	8	12	16	20	24	28	32	36	40
5	5	10	15	20	25	30	35	40	45	50
6	6	12	18	24	30	36	42	48	54	60
7	7	14	21	28	35	42	49	56	63	70
8	8	16	24	32	40	48	56	64	72	80
9	9	18	27	36	45	54	63	72	81	90
10	10	20	30	40	50	60	70	80	90	100

In fact, the whole task can be instantly halved if we spot that the table is diagonally symmetrical (compare either side of the diagonal line). This means we only have to learn half the facts, as the other half are mirror images. For example, $4 \times 7 = 28$ and $7 \times 4 = 28$ as well: we don't have to learn both facts. Add to this the fact that the 1 times and 10 times tables are very easy patterns to learn, and suddenly there are only about 20 facts left.

The wonder of tables

The grid also reveals other number patterns. What do you notice if you shade in the odd numbers? Why does this happen? Look at the numbers lying on the diagonal marked by the arrow – does the shape of the grid help to explain why these are called "square" numbers?

More surprisingly, the grid is an instant "fraction converter". Look at any two adjacent numbers within the grid – for example, 4 and 6 – and think of them as a fraction: $4/6$. Read along the columns or rows of these two numbers in either direction, and each number pair is equivalent – in the case of $4/6$, from $2/3$ to $16/24$ to $20/30$. The grid shows instantly, for example, that $24/32$ is the same as $3/4$ or that $35/40$ is $7/8$. This works both horizontally and vertically.

"Mathematics seems to endow one with something like a new sense."

Charles Darwin (1809–1882)

You may need to read through this chapter a couple of times before a lot of the ideas contained in it take root. Even so, have a go at the following questions to see how much progress you've made. There isn't a test at the end of every chapter, though there will be numerous invitations and challenges throughout the book to help you apply what you're learning.

1   In your head (no calculator):
    a: multiply 34 by 99
    b: multiply 3.99 by 7

2   a: What is $150 \div \frac{1}{2}$?
    b: What is $200 \div \frac{1}{5}$?

3   Odd numbers have an interesting pattern when we add them:
    The sum of the first two odd numbers is: $1 + 3 = 4$ (which is 2×2).
    The sum of the first three odd numbers is: $1 + 3 + 5 = 9$ (which is 3×3).
    What would the first 100 odd numbers add up to?

4   16 reversed is 61, and 28 reversed is 82. But in what circumstances could 16 and 61 actually be equal (and 28

and 82 also be equal)? (Clue: if you think this sounds familiar, you're getting "warmer"!)

5 a: What is the 100th term in this sequence? 7, 11, 15, 19, 23, 27…
   b: Here's another sequence: 2, 5, 8, 11, 14. How many terms of this sequence would it take to get to 146?

6 "Chain" style speed tests are popular in some newspapers. How quickly can you get to the ends of the following "chains"?

START 13	× 1	+ 1	square root of this	× 99	10% of this	double	ANSWER

START 340	halve	halve again	1/5 of this	× 4	× 5	ANSWER

7 What 5-digit number is missing from the highlighted section of this grid, and why?

		2	5	
4	1		2	3
	5	4		1
5		3		2

CHAPTER 2

CLEAR-HEADED CALCULATING

One of the potential confusions arising from the way we learn mathematics is that the methods we're taught to use on paper are very different from the ones that we're taught to do in our heads. This is not necessarily a problem, but often the difficulty lies in imagining we need to do a complicated calculation to solve a particular problem, when there might very well be a much easier shortcut that we can use. This chapter will help you to feel more at ease with manipulating numbers and introduce you to some arithmetical wizardry that bypasses long-winded calculation.

ALL IN YOUR HEAD

A few simple tricks and techniques will help you to become
fluent in mental calculations. Work through these slowly,
and you'll notice a significant improvement both in your
confidence with numbers and in your accuracy and speed.

DOUBLING AND HALVING

Want to work out how much two, or four, or eight cans of dog food will
be if you know the price per can? Or how much a dress will cost if it's
reduced by 50 per cent? Doubling or halving a number quickly in your
head is the basis for much mental dexterity with numbers.

Doubling

The secret to doubling a number is the ability to break it into separate
parts. You may do this in whatever way you find the most convenient.
For example, to double £3.47, you might choose to double £3, then
40p, then 7p, giving £6 + 80p + 14p or £6.94. Or you might prefer to
think of the original amount as £3.50 minus 3p, so you could double
the £3.50 to £7 and then subtract two lots of 3p to leave £6.94. This
latter technique is particularly useful in stores that insist on charging
£3.99 instead of £4 for items, hoping that we'll think of them as £3!

Doubling techniques are useful for large numbers, too, as
doubling 3,000 is no harder than doubling 3. To double 3,462, for
example, I would think of 6,000, 800 and 124 (remember that you're
free to partition in any way), giving 6,924.

Practise this every time you see a number over the next few days: simply double it. And double it again. Remember to look out for the neatest route: you can double 26 by doubling 25 (50) + 2, for example.

Halving

Halving requires a similar, though not identical, technique, and one I have successfully tried with children as young as eight years old.

Picture a bold horizontal line floating at eye level and arm's length in front of you. Picture the number you wish to halve sitting above the line. Partition the number into two or three pieces – whichever you feel is manageable. Place these pieces above the line. Now mentally halve each piece, and place the half below the line. Remembering what you placed where, push everything below the line together and add it up. Here's how it would work to halve 335

300		30		5
150	+	15	+	2.5

TOTAL: 167.5

Practise this over the next week or so and you'll be amazed how quickly you develop a more rapid fluency.

REAPING THE BENEFITS

Once you can double or halve with confidence, you'll notice that other, seemingly unrelated arithmetical tasks suddenly start to become easier. For example, by repeatedly doubling you can quickly find 4, 8, 16 and 32 times any number. Having eight people for dinner and the recipe allows 140g of pasta per person? No problem; just double, double and double again (280g, 560g, 1,020g). Similarly, by repeating the halving technique you can find $1/4$, $1/8$, $1/16$, and so on.

You can also use these techniques to calculate other quantities mentally. If you suddenly find that you've miscalculated and there are only six of you for dinner, think of this as 2 plus 4. So: double (280g) for × 2, double again (560g) for × 4, and add: 280 + 560 = 840g.

ABOUT A HUNDRED

Imagine that you're shopping in New York. So far, you've bought three items. They cost $1.99, $2.97 and $4.99 respectively. How do you work out the total? You could use a traditional column method:

$$\begin{array}{r} \$1.99 \\ \$2.98 \\ \underline{\$4.99} \\ \$9.96 \end{array}$$

More efficient would be to round these figures: they are approximately $2, $3, and $5 respectively, giving a total of $10. Then add the small amounts by which the actual figures fall short of the rounded-up figures: 1¢ + 2¢ + 1¢ = 4¢. Subtract from the total: $10 − 4¢ = $9.96.

This has the added benefit of being faster than either the written method or resorting to a calculator. See how quickly you can add up three items costing $3.99, $4.97 and $2.98 ...

LOOKING FOR FAIRINGS

To add several numbers in your head, look for pairs that total 10 (or 100). For example, how might you add up: $20 + 40 + 32 + 80 + 60$? You could plod along the line:

$20 + 40 = 60$

$60 + 32 = 92$

$92 + 80 = $ er ...?

Instead, think smart, and look for pairings to make the task simpler:

$20 + 80 = 100$

So does $40 + 60$

So that's 200, with just 32 to add on at the end: 232

Take this method a step further by looking for numbers that are almost round numbers. To add up 41, 65 and 59, imagine removing 1 from the 41 and giving it to the 59; you now have 40 and 60, which is easier to add. Very quickly, you can then reach the correct total: 165.

Use this idea to add 23, 34, 56 and 27 in your head. (Remember to add them in any order you find helpful.)

BIG, SCARY NUMBERS

A popular misconception is that working with large numbers is hard. This nonsense is easily dispelled by the following example. Which of these numbers can you multiply by 6 in your head most easily: 0.37, 134, or 1,000,000? Don't let your mind be fazed by a long string of digits: look at what the number actually is, and see if you can simplify it.

ELEVENS LIKE LIGHTNING

Nobody would be surprised if you could multiply a number, even a large number, by 10 or by 100 faster than a calculator. But what about multiplying by 11? Here's how.

* * * * *

What's the quickest way to multiply 23 by 11? One shortcut would be to multiply by 10 and then add 23, but even quicker is simply to add the two digits and put the answer in the middle:

$2 + 3 = 5$, so $23 \times 11 = 253$.

If the sum of the two digits comes to more than 9, carry the 1 over and add it to the left-hand number:

84×11: 8 (12) 4 = 924

Try magically multiplying: 34×11, 52×11, 69×11, 85×11, 93×11.

This trick works with larger numbers, too:

4623×11

4   (10)   8   5   3 = 50853

Bigger numbers take more practice, as you have to hold each addition in your head *and* remember to carry over a 1 each time a pair adds up to 10 or more. But it's a great trick for mental dexterity, and with practice you can multiply even thousands in your head faster than a calculator.

DEVELOPING YOUR MENTAL MUSCLES

Number games are a good way to hone the skills you've begun to acquire. Here are a few to get you going, but look out for similar ones in newspapers, puzzle books and other popular sources. The first stage is to understand the principle and work your way to the right answer, by the shortest possible route; the second stage is to increase your speed without sacrificing accuracy.

QUICK CALCULATION CHAINS *(solutions on page 133)*

The two "chain" puzzles below are more complex than those you've already encountered on page 31 – but try to complete both of them in your head. Halving, doubling and rounding numbers can all help. How fast can you reach the end to find the correct answer?

START WITH 20	find 10%	× 2	× 12	− 34	÷ 1/3	1/7 of this	find square it	× 99	ANSWER

START WITH 7	square it	× 11	− 439	square root of this	find 20%	cube it	square it	+ 6	ANSWER

PATTERN SPOTTING *(solution on page 133)*

For each of the diagrams below, work out the hidden relationship between the numbers, then use it to fill in the missing boxes.

A

| 4 | 3 | 10 |

| 1 | 8 | 6 |

| 2 | 5 | ? |

B

| 2 | 3 | 10 |

| 4 | 3 | 14 |

| ? | 3 | 12 |

C

| 1 | 4 | 5 |

| 6 | 3 | 39 |

| 4 | 3 | ? |

CROSSED LINES

(solution on page 133)

Place the numbers 1 to 9 into the empty circles so that each line of connected circles adds up to the value at the end of the line. No number should be used in more than one circle.

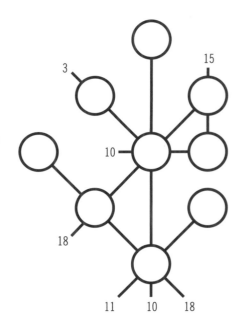

KAKURO *(solution on page 133)*

You may already be familiar with Kakuro – this puzzle is a relative of Sudoku, but unlike in Sudoku, you need mental addition and subtraction, rather than logic alone, to help you toward the solution.

To solve Kakuro, you must place a digit from 1 to 9 in each white square. The total of each run of numbers must equal the total given to the left or top of it. *No number can be used more than once in any run.* For example, two of the blank squares on the bottom row require their numbers to add up to 11. And the first number will contribute to the 24 total indicated by the "down" clue just above.

A possible start would be the pair of squares marked with stars. Only 7 + 9 will give the required total of 16 (as 8 + 8 is not allowed), but which way round: 79 or 97? The former would add 9 to the column of four squares that must total 13. But 13 − 9 = 4, and you can't make 4 with three different numbers. Now complete the grid.

Do Kakuro regularly and you'll find you become more adept at mental arithmetic and at recognizing number combinations, such as 22 − 6 = (7 + 9), or 34 − (8 + 9) = (4 + 6 + 7).

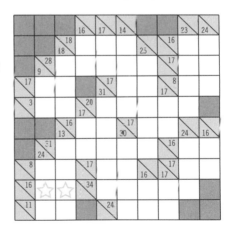

THE ELECTRONIC BRAIN

Calculators are great – where would we be without them? However, although they have their strengths, they also have their weaknesses. How do we decide when to use them, or even whether we should trust what they tell us?

ARE CALCULATORS ALWAYS RIGHT?

Often, using a calculator is the most efficient way to work something out. Try using pencil and paper to find the cost of 23,587 barrels of oil at $126.16 per barrel, for example. However, there is a danger inherent with calculator use. I refer to it as "mindless deference syndrome". More basically, "It's a machine, so it must be right." Not at all! You are far cleverer than a calculator. Consider the following.

The families in Decimal Drive are surveyed about the number of cars they own. One family has one car; the other nine have two each. You might instantly see that this is a total of 19 cars. But ask your calculator to work out $1 + 2 \times 9$. Does it give you an answer of 27? Clearly wrong. Scientific calculators know that we need to do the multiplication before the addition, but an ordinary calculator simply tackles each task in the order it is given, so it will work out $1 + 2$ and then multiply by 9. What you knew, perhaps without even realizing it, was that you needed to multiply 2 by 9 first, and *then* add 1.

With a simple calculation the error is obvious, but it's very easy to make mistakes like this, both with and without a calculator, when faced with a chain of calculations, such as adding on or taking off percentages or fractions (see page 52). Think about the order in which you need to approach the following problems:

- How would you work out the price of a piece of electrical equipment *without* tax, if you knew the price once a sales tax of 15% had been added on?
- Your boss offers you a 10% pay cut but promises to give an 11% pay rise immediately afterwards. Tempting? (In fact, this results in a 0.1% *drop* in salary rather than a 1% increase. Can you see why? If you need a clue, think 10% of *what*, and 11% of *what*?)

THERE AND BACK AGAIN

Here's another intriguing trick to try with a calculator. Type in a three-digit number, and then type the same three digits again in the same order, so that you have a six-digit number — for example, 237,237.

Press '÷ 11 ='. You should now have a five-digit number. Keeping this answer in the display, press "÷ 13 =". You should now have a random three- or four-digit number. Press "÷ 7 =". You will now be staring at the three-digit number you started with!

How does this work? The clue is that $7 \times 11 \times 13 = 1,001$. To multiply a number by 1,000 you just add three noughts, so multiplying by 1,001 results in the original three-digit number being repeated:

$$237 \times 1000 = 237,000$$
$$237 \times 1 = 237$$
$$237,000 + 237 = 237,237$$

Division, as you were doing with your own three-digit number, simply "undoes" the multiplication. Because multiplication and division undo each other's work, they are known as "inverse operations".

MAGIC MATHS: 6801

There are a number of apparently magical mind-reading tricks that begin "Think of a number…". How on earth do they work, and can you do them, too? Here's how you can amaze your friends with your mathematical dexterity.

THE PUZZLE
Try the following trick.
- Write down any three-digit number, making sure that all three of the digits are different.
- Now write down the number again, but backwards (ie if you had chosen 259, write 952).
- Subtract the smaller number from the larger one. Your answer should be a 3-digit number. However, if it is a 2-digit number, simply add a zero in front of it (ie for 37, write "037").
- Reverse the answer again, and this time add instead of subtract. You should now have a 4-digit number.
- Now for the magic bit. Turn this page upside-down and look at the title: you should now see your answer staring at you!

THE REASON WHY
To unpick how this trick works, think of what you have done in terms of the "hundreds, tens and units" columns of school arithmetic. Suppose the number you first thought of was 348; reversing it would have given you 843. Your first move was:

$$\begin{array}{r} h\ t\ u \\ 8\ 4\ 3 \\ -\ 3\ 4\ 8 \\ \hline 4\ 9\ 5 \end{array}$$

Whichever number you chose, two facts remain constant: in the units (u) column the lower digit will always be greater than the upper digit (think about it!), and in the tens (t) column the numbers will be the same – in this case, 4. Whichever way you were taught to subtract, you'll find that the "borrowing" from the t column in the course of subtracting the u digits means that the answer in the t column will always be 9. Crucially, the h and u digits of your answer will also always add up to 9. The significance of this becomes clear when you move on to the second stage:

$$\begin{array}{r} h\ t\ u \\ 4\ 9\ 5 \\ +\ 5\ 9\ 4 \\ \hline \end{array}$$

The digits in the u column will always add up to 9, as will those in the h column (since they are the same in reverse). And the t column always contains two 9s, making 18:

9 18 9

Of course, the 1 of the 18 is carried over and added to the h column, making the final total:

10 8 9

CONVERSION BENCHMARKS

When converting currencies or measurements, memorable "benchmarks" can make quick mental calculation a lot easier – for example, if you need to figure out whether that vacation souvenir is good value, or want to know what a distance given in kilometres means in miles ... or vice versa.

DO I NEED TO KNOW EXACTLY?

Sometimes, precise conversion is important: you wouldn't want the bank to use a rough estimate to work out the exchange rate for your vacation currency, for example, or the pharmacist to guess at a prescription dosage. But often what's most helpful is to have at your fingertips some realistic, easy-to-remember equivalents so you can quickly reach a "ballpark figure".

There is more on the value of estimating in Chapter 4 (see page 88), but here are some ideas for helpful benchmarks to use when you need to do quick conversions in your head. Create your own for whatever circumstances you encounter, whether it's converting recipe amounts, working out how far you have to travel or envisaging the floor area of houses for sale abroad.

HOT OR COLD?

The Celsius and Fahrenheit scales hit the same reading at -40°. This is of certain academic interest, and useful general knowledge for a quiz, perhaps, but how about a more practical guide to what the weather is like? You probably already know that 0°C = 32°F, and here are two "reverse number" benchmarks:

$$16°C = 61°F$$
$$28°C = 82°F$$

Commit these to memory, and you'll find it easy to place other readings in relation to them, without having to do tricky conversions.

MONEY, MONEY, MONEY

Rather than trying to remember exact exchange rates, provide yourself with a few comfortable benchmarks. If 1 zloty is worth 23¢, then there are *roughly* 4 zlotys to the euro, and €50 is a bit over 200 zlotys. Keep this in mind and it will easier to judge whether a 6-zloty cup of coffee in Warsaw is a rip-off or not and if you can afford that 600-zloty dress.

A CURIOUS COINCIDENCE

A basic benchmark between miles and kilometres is that 5 miles = 8km. However, from 5 miles onward, the mile/km equivalents also equate with surprising accuracy to the Fibonacci sequence (see next page). So 21 miles is around 34km, 34 miles around 55km, and so on. Make yourself a pair of Fibonacci number strips as shown below, place them one over the other, and you have an instant conversion table.

km	8	13	21	34	55	89	144	233	377	610	987
miles	5	8	13	21	34	55	89	144	233	377	610

FIBONACCI

Leonardo of Pisa (*c*.1170−*c*.1250), known as Fibonacci, was an immensely talented mathematician, and the arithmetical sequence he discovered in 1202 can be detected in an amazing variety of places – some very unexpected.

LEONARDO AND THE RABBITS

Fibonacci discovered the sequence that bears his name while investigating how fast rabbits could breed, and this rabbit problem is often used even today to introduce the sequence to new audiences.

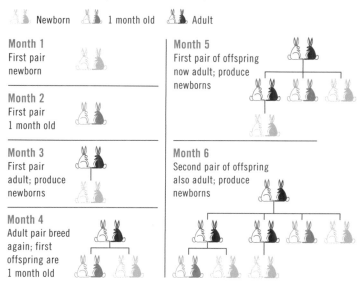

Newborn   1 month old   Adult

Month 1
First pair
newborn

Month 2
First pair
1 month old

Month 3
First pair
adult; produce
newborns

Month 4
Adult pair breed
again; first
offspring are
1 month old

Month 5
First pair of offspring
now adult; produce
newborns

Month 6
Second pair of offspring
also adult; produce
newborns

If a pair of rabbits bred once a month, starting at 1 month old and always producing a pair (1 male, 1 female) a month later, how many pairs would there be after one year, assuming none died? Fibonacci found that after one month there'd still be 1 pair; after two months, there'd be 2 pairs (1 adult, 1 newborn) after three months, 3 pairs (1 adult, 1 month-old, 1 newborn), but from the fourth month the pace picks up: 5 pairs, then 8, 13, 21 and so on. A notable feature of this sequence is that each term is the sum of the two previous terms:

1 1 2 3 5 8 13 21 34 55 89 144 ...

THE SEQUENCE IN NATURE

The rabbit breeding pattern is obviously an unrealistic example, but the Fibonacci series abounds in nature. The head of a sunflower is a celebrated example. Its seeds grow in spirals formed in Fibonacci numbers: for instance, the 21st seed, counting clockwise, will sit next to the 34th seed in the next anti-clockwise spiral. The outer rim might typically contain 88 seeds one way and 55 the other. Pine cones also have spirals formed in Fibonacci patterns, as do many types of seashells. The petal count of many flowers, including buttercups, wild roses, corn marigolds, chicory and michaelmas daisies, conforms to a number in the Fibonacci sequence: 5, 8, 13, 21, 34, 55 or 89 petals.

This sequence is an example of the fascinating relationship that exists between numbers. As we saw on page 47, it also equates to the relationship between miles and kilometres. The link between the two measures is the golden ratio, which we will explore in the next chapter.

NUMBER RELATIONSHIPS

People of different nationalities can all say the same thing but sound completely different, each saying it in his or her own language. Mathematics has been described as a universal language, but it too can be used to say the same thing in a variety of different ways. You may recognize that 50 per cent and 0.5 are just different ways of saying "half", but in this chapter, you'll gain a better understanding of fractional numbers; and you'll also learn, for example, how 2 can also be 10 and even a or x.

Appreciating the different ways in which amounts can be expressed, and the relationships numbers have with each other, will take you a major step forward in becoming relaxed about using them in everyday life.

PARTS OF A WHOLE

Fractions, decimals and percentages are often confused, but this need not happen. Far from being separate entities, they are simply different ways of expressing the same thing: parts of a whole. Choosing to work in percentages or decimals can make some calculations easier, but the important thing to remember is that they are just fractions in different form.

A FRACTION BY ANY OTHER NAME

Fractions immediately show us how many equal parts a particular whole is divided into: $1/16$ is one of 16 equal parts; $3/4$ is three out of four equal parts; and so on. Because our number system is based on 10 (see page 75), tenths and hundredths are very common divisions, and these are often more conveniently expressed as decimals and percentages: $1/10$ is 0.1 or 10%, and $1/100 = 0.01$ or 1%.

Marking off equal divisions along a line (representing the whole, or 1) shows how the same fractions can be differently expressed.

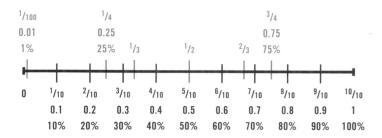

WHY NOT ALWAYS USE THE SAME FORMAT?

We commonly talk and think in fractions, perhaps because they can be so clearly visualized — we can easily imagine half an apple, or dividing a cake into eight equal portions. But percentages are a more convenient format for describing how a total has been divided up into *unequal* parts. Suppose a store's profits, broken down by department, were expressed as fractions: leisurewear: $1/4$; children's toys: $3/20$; sports equipment: $2/5$; menswear $1/5$: It's not immediately obvious which is the most profitable department. However, convert these into percentages (25%, 15%, 40%, 20%) and all will become clear.

Adding and subtracting are usually easier in decimals than in fractions. Which is easier, $1.125 + 3.4$ or $11/8 + 32/5$? This is one reason why most currencies are now decimal-based. Percentages and decimals also often make it clearer what a particular fraction actually represents. Imagine trying to visualize $^{14}/_{25}$ of a number, for example. It is far easier to consider this as 56% ($^{56}/_{100}$), or just over half.

PERCENTAGES AND PERCENTAGE CHANGE

Percentages are also a convenient way of conveying increases or decreases such as pay rises or the mark-up on retail goods. Such comparisons show the **percentage change**.

If a child has grown from 80 to 100cm in a year, the difference is 20cm. To find the percentage increase this represents, simply divide the amount of change by the starting amount, and then multiply this number by 100. As $20 \div 80 \times 100 = 25$, we can see that there has been a 25% increase in the child's height.

Talking in terms of percentage changes rather than actual figures throws up some interesting possibilities, as journalists and politicians have been quick to perceive. Suppose that last year, of 1,000 heart operations carried out, only one resulted in the patient dying on the table. This year, in a similar time period, just two people didn't survive. Not a bad surgical performance by any standards, but imagine the headline:

HEART OP DEATHS INCREASE BY 100%!

True, but not a helpful interpretation of the figures. Examples in the media of percentages being used in a misleading fashion are not hard to find. Watch out for instances where percentage change has been used for effect, and look behind the figures to get a better picture of what is really going on.

Compare and contrast ...

Here is a way in which percentage change *can* be helpful in influencing the way we perceive relative amounts. Suppose, on your way to buy a new car, you pass a store and go in to buy your favourite chocolate bar. You're disgusted to find that it has gone up from 40¢ to $1! Indignant, you make a 10-minute detour to buy your chocolate bar in another store for the usual price. A little further on you notice a huge billboard advertising the same model of car that you have been planning to buy for $12,000, but offering it at a garage half a mile away for $11,999.40.

It wouldn't occur to you to detour to the other garage to buy the car – who would do that, just to save 60¢? But that's exactly what you just did with the chocolate! What makes the two situations seem so different is the percentage change: the chocolate represented a massive 150% price increase, but the cheaper car was a discount of 0.005% – just five-thousandths of 1%.

PERCENTAGE POINTS

A common error is to confuse percentage points with an actual percentage. If a financial newspaper reports, for example, a rise of 3 percentage points in overdraft rates, it is saying that there has been a rise in the rate of three points: from 6 to 9% perhaps, or 22 to 25%. This is very different from a 3% rise.

EASY CONVERSION

Here is a simple reminder of how to translate numbers from one form to another. These are not necessarily the quickest methods, but they always work – there are no tricky exceptions.

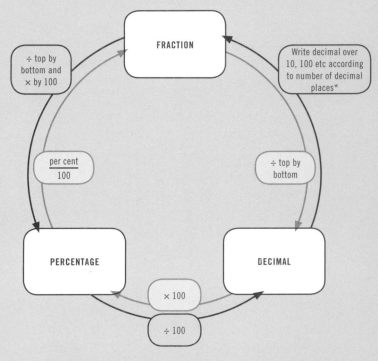

*so, for example, 6.2 would be $6^2/_{10}$ and 6.25 would be $6^{25}/_{100}$. Fractions can then be expressed more simply – in these cases, as $6^1/_5$ and $6^1/_4$.

MENTALLY CALCULATING PERCENTAGES

The key to working out percentages in your head is to break them down into blocks that are easy to remember. It helps to use the number patterns and the skills in halving and doubling that you've already learnt (see pages 34–36 if you need a recap).

Basic building blocks

The most basic percentage block is 10%. To find 10% of any number, you simply need to divide it by 10, by moving every digit one place to the right (or, to put it another way, by moving the decimal point one place to the left).

- To find 10% of 26: $26 \div 10 = 2.6$
- 10% of $13.7 = 1.37$
- 10% of $12,340 = 1,234$

  (1,234.0, strictly speaking, but we can dispense with the zero)

To find 1%, just do the 10% trick twice:

- 1% of $26 = 0.26$
- 1% of $6,153 = 61.53$
- 1% of $13.7 = 0.137$

Halving and doubling helps to make a number of other percentages easy to calculate. 50% is half, of course, and you can work out 5% quickly either by finding 10% and then halving, or by finding 10% of 50%, as in the following examples:

- 5% of 26: 10% = 2.6; half again is 1.3
- 50% = 13 and 10% of 13 = 1.3

By thinking in combinations of these basic blocks you can very easily find any percentage. For example, what is 60% of 3,000km?

- A good route is: 50% + 10%: 1500 + 300 = 1,800km. Done!
- An alternative would be to find 10% (300km) and then multiply this by 6.

Quantities such as 19% require a bit more imagination, but need not be daunting. How about:

- 20% (10% doubled), then subtract 1%?

Here's some good news. At school we were often told that we must do calculations in one specific way, but as your confidence grows you'll realize that there are various ways to work things out, and you can choose the method that makes the most sense to you. Calculating percentages in your head is a good example of this flexibility. To find 16% of something, you might choose to find 1%, then double it four times in a row. I might prefer to find 10%, add another half of that (5%), and add another 1%. A third person might choose to find 10%, double, then subtract 4 lots of 1%. And we would all be correct!

How would you find:

- 75% of a number?
- 48%?
- 17.5%?

Practise whenever you see the opportunity – sales discount offers of 25% or 33%, for example, or a 15% service charge.

RATIOS AND PROPORTIONS

Ratios and proportions are another way to express parts of a whole, but they concern relative rather than actual amounts. Mixing paint, mixing cocktails and designing aesthetically attractive buildings all involve using the right ratios to achieve pleasing proportions.

WHO USES RATIOS?

A ratio, written as X:Y, describes one amount in relation to another. You may remember learning to calculate ratios at school – questions about how many tins of blue and yellow paint you need for a particular shade of green have long been popular in mathematical examinations – but ratios are widely used in the real world, too. Chefs, anaesthetists, builders, hairdressers, to name but a few, all need to have a good appreciation of ratios if they are to mix their hollandaise, anaesthetic, concrete or hair dye successfully.

Look around your home and you'll find ratios as instructions or information about mixed ingredients: "Dilute 1 part of squash with 4 parts of water"; "For every cup of rice add twice the volume of water"; "The NPK fertilizer values in this pack are 8:2:6."

Television screens come in a huge range of *sizes*, but a very limited number of *ratios*, so that, whatever the width, the height is in proportion. A widescreen 16:9 TV is 16 units of width for every 9 units of height, irrespective of size. And when manipulating digital photos on a computer, you'll probably have an option to maintain the ratio of the picture's dimensions, so that if you adjust the width of the photo, the height automatically adjusts "in pro", or in proportion.

WORKING WITH RATIOS

Many of us have experienced sauces that are too runny, concrete that won't set, and so on, because the ratio of ingredients is wrong. A hairdresser notes that she must mix hydrogen peroxide and red colorant in the ratio 4:1: ie if she uses 40ml of peroxide, she needs to add 10ml of colorant. Accidentally, she pours 20ml of colorant into the mixing bowl. Thinking quickly, she decides to balance the extra 10ml of colorant with an extra 10ml of peroxide. Soon, however, she may hear the screams of the newest, reddest head in town ...

Of course, she should have added another *40ml* of peroxide. To keep ratios constant, the numbers concerned must all be **multiplied or divided** by the same amount: unless the ratio is 1:1, adding or subtracting the same amount of both will change the ratio. By adding only 50ml of hydrogen peroxide, the mixture's ratio became 50:20. Simplified to its lowest figures, this is only 5:2, or 2.5:1.

PROPORTIONS

These are linked with ratios, but are slightly different. Ratios show the relationship of one part to another part; proportion expresses one part in relation to the whole. The ratio of the hapless hairdresser's correct mix was 4:1, but this could also have been described as 1 part of colorant to 4 parts of peroxide. This makes 5 parts altogether, so the *proportion* of colorant is one fifth (of the whole mixture).

USING RATIOS TO UNDERSTAND LARGE NUMBERS

Thinking in terms of ratios can be useful when trying to focus on what large figures really mean. If the population of a fictitious country is estimated to be 60 million, and its spending on a national health programme is 75 billion, then we can use the ratio between these figures to get an idea of how much is spent per capita:

60 million:75 billion

Reducing both sides of the ratio by stages gives:

$60:75,000 = 6:7,500 = 1:1,250.$

So, on average, 1,250 is spent on each member of the population.

FIGURING WITH RATIOS *(solutions on page 134)*

Try your hand at working with ratios yourself.

- Reverend Chalkey has bats in his belfry. He notices that there are 30 brown bats for every 20 black bats. If there are 200 bats in total, how many of them are brown?
- A bag of 90 sweets is to be shared between William, Thomas and Rebecca in the ratio of their ages. William is 5, Thomas 6 and Rebecca 7. How many sweets will they each receive? (Hint: try to reduce the ratios and/or find out what one share is worth.)
- Blood is a mix of plasma and various kinds of cells. In a healthy human, these cells make up 45% of blood. How would you

describe the proportion of plasma? And what is the ratio of plasma to cells, reduced to the simplest ratio?

Before you read any further, measure the horizontal and vertical sides of a credit card as accurately as possible. Use a calculator to divide the larger measurement by the smaller. You will probably get an answer of around 1.6, meaning that the ratio of the two dimensions is 1.6:1. This is no accident – turn to page 66 to find out why.

A SPECIAL RELATIONSHIP: PI

Pi (usually written as the Greek letter π) is one of those legendary numbers that has fascinated mathematicians for centuries. It is perhaps the best-known example of an irrational number, meaning it can never be written as an exact decimal.

Despite being an "irrational" number – impossible to write down precisely in decimal form – π represents a very precise relationship: that of a circle's diameter to its circumference. The ancient Greeks discovered that dividing the length of the perimeter of a circle (the circumference) by the length of any straight line cutting the same circle in half (the diameter) would always give the same answer: slightly more than 3.14. A fun way to show this is to get a group of people holding hands in a circle, then announce that you'll need exactly a third of that number to make a straight line that reaches across the circle. You'll be right every time!

The Greeks also worked out that you could find the area of any circle by squaring its radius and multiplying the answer by π. (You may remember the two formulae from school: A[rea] $= \pi r^2$, and C[ircumference] $= \pi d$ (or $2\pi r$).)

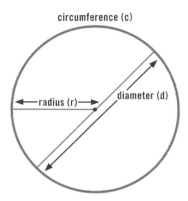

circumference (c)

radius (r)

diameter (d)

The number 3.14 is only an approximation of π. To 25 decimal places, π equals 3.1415926535897932384626433 – but even this is not the exact value. Computers have calculated π to more than a trillion decimal places and still not reached the end.

A LITTLE HELP FROM EINSTEIN

If you'd like to impress your friends by memorizing the first 15 digits of π, here's a quote from Albert Einstein that will help you:

"How I need a drink. Alcoholic, of course, after the heavy chapters involving quantum mechanics."

How does that help? Try counting the letters in each word.

THE GOLDEN RATIO

Originally represented by the Greek letter Φ (phi), the golden ratio is also known as the golden section, or even the divine ratio, owing to its perceived beauty. It was discovered by the Greek mathematician Euclid, but much of what is commonly thought to be true about it is at best tenuous, and at worst completely wrong!

WHAT IS THE GOLDEN RATIO?

As we saw on page 59, a ratio is a comparison or relationship between two quantities. Consider a line, AB, with a point C on the line but slightly off-centre, so that the ratio of the distances AC:CB is equal to the ratio AB:AC. This is the golden ratio.

A   C   B

The exact value of the golden ratio can be found by halving one more than the square root of 5 or, to write this mathematically, $(1 + \sqrt{5}) \div 2$. The ratio is an irrational number – this means that it cannot be exactly expressed as a decimal, and has no recurring digits (see page 53). This quality makes it important in nature (see page 67) as well as in art. Before we look at occurrences of the golden ratio, however, it is interesting to debunk a couple of myths that have grown up surrounding it.

TWO COMMON FALLACIES

Many people believe that the golden ratio is exactly 1.618, including the author Dan Brown who cites this number in *The Da Vinci Code*. Although 1.618 is a good approximation, it is not exact, since the golden ratio is an irrational number.

It is also widely held that ancient civilizations such as the Greeks, the Romans and the Egyptians were obsessed with the golden ratio and used it extensively in their buildings, as did many Renaissance artists. This is a slightly misleading claim. The golden ratio is undoubtedly pleasing to the eye in its proportions, and therefore it's not surprising that many buildings or paintings that are considered aesthetically attractive contain close approximations to it (for example, the height and width of the Mona Lisa's face perfectly fit the golden ratio). However, most instances of this kind tend to fit the evidence to the theory retrospectively – artists and architects created proportions that pleased the eye, and these often turned out to accord with the golden ratio. The phrase "golden ratio" itself was not used until 1835!

"The beautiful seems right by force of beauty . . ."

Elizabeth Barrett Browning (1806–1861)

THE GOLDEN RATIO AND THE FIBONACCI SEQUENCE

The ratio between adjacent terms in the Fibonacci sequence of numbers (see page 48), found by dividing the larger by the smaller, rapidly tends toward the golden ratio:

$$1 \div 1 = 1$$
$$2 \div 1 = 2$$
$$3 \div 2 = 1.5$$
$$5 \div 3 = 1.6666$$
$$8 \div 5 = 1.6$$
$$13 \div 8 = 1.625$$
$$21 \div 13 = 1.615$$
$$34 \div 21 = 1.619$$
$$55 \div 34 = 1.618 \qquad (\text{and } 34 \div 55 = 0.618)$$

(You may also notice that this explains the kilometre/mile conversion table on page 47: since 1 mile is approximately 1.6km, the ratio between them approximates to the golden ratio and so closely shadows the Fibonacci sequence.)

Islamic geometric art makes extensive use of Fibonacci numbers to generate some exquisite patterns. The artists considered the ratios of successive terms to be particularly sacred, since $1/0.618 = 1 + 0.618$, and 1 is considered to be unity (or the universe).

The happy visual balance of a rectangle in the golden ratio has led to its use in many modern designs, too, including televisions, windows, doorframes, magazines ... and credit cards.

THE GOLDEN
RATIO IN NATURE

The pleasing proportions of the golden ratio explain its recurrent use in art and design, but it is the frequent natural occurrences of this ratio and the related Fibonacci sequence that are truly fascinating. One outstanding example is the angles at which leaves grow.

Ideally, as a plant grows, its new leaves should emerge at a slightly different angle from those below them, so that new growth doesn't exclude light from existing foliage. But what is the best angle to turn? A turn of 90°, for example, would mean that the fifth leaf would precisely obscure the first. Nature uses the golden ratio to solve the problem. On many plants, each leaf grows (approximately) 1.618 turns around the stem. This works out at one complete turn (360°) plus about 222.5°. Because this angle is tied to the golden ratio, which is irrational and therefore never precisely repeats, there is never an exact overlap, no matter how large the plant grows.

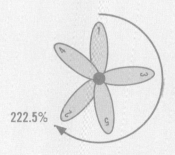

222.5%

MEANINGFUL AVERAGES

We all know about averages: they are the middle ground, the halfway measure. And yet they can be calculated in more than one way, giving different results, and can easily be misleading and misused.

MIS-STATING THE OBVIOUS

Consider the following statements:

"Half the nation's children are performing below average in mathematics!" (headline in *The Times*)

"The average family has 2.3 children."

"Most people have more than the average number of legs."

These are all technically true, yet none is terribly helpful, and the last one just makes us laugh. Essentially there are three quite distinct ways of finding an average, and each is appropriate in the right circumstances. Using the wrong method leads to silly or misleading results like those given above.

MEANS, MEDIANS AND MODES

The most common type of average is the **mean average**, whereby all the data are simply added up and divided by the total number of items. For example, the mean average of 2, 6 and 7 is 5, since $2 + 6 + 7 = 15$, and $15 \div 3 = 5$. But this doesn't tell us about the individual numbers. For example, the mean average of -82, 0, 100 and 2 is also 5. It is the use of the mean average that makes the statement about 2.3 children true mathematically but not, of course, realistic.

An alternative is to work out the **median average**. This is useful if we wish to discover, for example, what a typical student might score in an exam. To work it out, scores are put in ascending order, and the median is simply the centre value. If there is an even number of scores or values, the median is halfway between the two centre values. Since half the numbers in the list will always be below this centre value, using the median average makes the statement from *The Times* true; however well children do, half of them will always be below the median.

Sometimes neither the mean nor the median is a useful measure, as we want to measure the average in the sense of the most common or popular item or occurrence. Suppose a market research company wished to know how many vacations families took each year. In surveying the 10 families who live in the fictitious Decimal Drive, they might get the following answers: 1, 1, 1, 1, 1, 2, 2, 3, 4, 5. What would best represent the average answer?

The mean would be $(1 + 1 + 1 + 1 + 1 + 2 + 2 + 3 + 4 + 5) \div 10 = 2.1$. The median would be 1.5 (halfway between the middle two numbers, 1 and 2). But both of these figures are misleading, as you can't have a fraction of a vacation! It makes more sense to look at the most common response, which is 1. This is known as the **mode**, or **modal average**. The French even have an expression, *à la mode*, meaning "popular" or "fashionable". In this case, because the most frequent answer is 1, we say that this is the mode: the modal average among families in Decimal Drive is one vacation a year.

Appropriateness should dictate which way you choose to express an average. The statement that most people have more than the average number of legs is only true of the mean or median, since the minority who have one or no legs bring down the average to below two – probably somewhere between 1.99 and 2. But the commonsense answer would be to talk about the modal average, which of course would be two legs.

"A good decision is based on knowledge and not on numbers."

Plato (*c.*428–347BC)

A IS FOR ALGEBRA

$E = mc^2$. $a^2 + b^2 = c^2$. Sound familiar? For many people,
algebra epitomizes all that once seemed daunting about
mathematics. it appeared pointless and abstract, and to
this day it is often a word laden with negative emotion.
Yet it need not be like this: algebra can be looked at in a
very different way indeed ...

WHAT EXACTLY IS ALGEBRA?

A typical lay definition would probably not go much beyond "letters
and equations". But this does not help us to understand the purpose
or indeed the value of algebra, so here is a slightly different definition:

**Algebra is a part of the language of maths that uses symbols to
represent numbers.**

This is not in itself a complicated idea. We use symbols all the time
to represent something else: barcodes, wiring diagrams, ticks,
crosses, arrows, so why not the letter a, b or x to represent a number?

AN ALL-PURPOSE SYSTEM

Algebra allows us to think in ways that are not limited to a particular
number or range of numbers. It enables us to make generalized
mathematical statements about how numbers will behave within
certain given parameters. For example, you've probably come across

tricks of the "Think of a number ..." type. They always seem to work, but how can people be so certain that they'll come out with the predicted answer? Here's a simple example.

Think of a number
Double it
Add 6
Divide by 2
Subtract the number you first thought of
Your answer is 3

But will the answer always be 3? How can we *prove* that this would work with any number? Are there any starting numbers for which it would not work? To find out, we can use some basic algebra.

Instead of starting with any specific number, let's start with an **x**. This letter can stand for any number we like.

- Doubling the **x** gives us 2**x** (2 times **x**, or "2 of the thing we are calling **x**").
- Adding 6 gives us 2**x** + 6. Note that we are not adding six of our number (which would be another 6**x**); we are specifically adding 6, as the puzzle asks.
- Next, we have to divide by two. This means halving each part of 2**x** + 6. Half of 2**x** is **x**, and half of 6 is 3, so half of 2**x** + 6 is **x** + 3.
- Finally, we need to subtract the number we started with, and this is the line that holds the key to the trick. We are simply taking away what we chose at the beginning – the number we are calling

x. Take **x** from **x** + 3 and we're left with just 3. Nothing we did made any reference to the size of the original number, so we have proved that this trick will work, *whatever* number **x** may stand for.

WHY IS ALGEBRA SO IMPORTANT?

The idea of making a symbol stand for any number at all is amazingly powerful, as it allows us to ask "what if?" about a whole range of circumstances or, as mathematicians would call them, variables. Algebra formed the basis for calculus, a massive step forward in mathematical thinking that opened up possibilities in fields such as engineering and physics. Without algebra we would never have been able to visit the moon, build jumbo jets or make computers.

One of the things that first brought C.F. Gauss (see page 24) into the public eye was his prediction of where astronomers should look for Ceres. When this "lost" minor planet was rediscovered, in 1801, it was exactly where Gauss's algebraic calculations had said it would be.

EQUATIONS

Equations are a much-maligned but crucial part of algebra. In their simplest form, they are merely statements that two or more expressions have the same value. Solving an equation involves finding out the value of symbols representing the unknown parts of these statements. Mathematicians often describe equations as *beautiful*. Although this idea may amuse (or bemuse) you, there's no denying that they contain the capacity to reveal truth, certainty and logic all at once, and many people find this very pleasing.

FERMAT'S LAST THEOREM

One of the most talked-about equations of the 20th century was Fermat's Last Theorem. The 17th-century French mathematician Pierre de Fermat (1601–1665) knew that it was possible to add two squares in order to produce another.

*   *   *   *   *

For example:

$$3^2 + 4^2 = 5^2$$
$$5^2 + 12^2 = 13^2$$

His theorem was that it was not possible to do this with anything but squares; not cubes (^{3}), nor fourth powers (^{4}) or anything higher. Expressed algebraically, Fermat surmised that if a and b were positive integers, then $a^n + b^n = c^n$ was an equation that had no solutions for any value of n higher than 2.

*   *   *

Proving it, however, became a challenge that confounded generations of mathematicians, and this simple-looking theorem remained unproven for more than 300 years, until the British mathematician Andrew Wiles solved it in 1995. Fermat himself tantalizingly wrote, "I have a truly marvellous proof of this proposition which this margin is too narrow to contain." We shall never know whether he was right!

Fermat's equation is a superb example of how simplicity and complexity can be encapsulated in a single, remarkable, beautiful statement of mathematical truth.

NUMBER BASES

Our Western number system is based on multiples of 10, and is known as "denary" or "base 10". This is something we mostly take for granted, but there is actually no limit to the number of possible bases we can use. In fact, we do work in other bases as well, sometimes without even realizing it.

WORKING IN DIFFERENT BASES

The addition sum below probably looks familiar from school, when you learned to start with the units on the right and "carry over" the 10s to the next column to the left (the 100s). The value of each column increases, from right to left, times 10 (units, tens, hundreds etc).

$$h \; t \; u$$
$$2 \; 5 \; 8$$
$$\underline{1 \; 7 \; 7}$$
$$\underline{4 \; 3 \; 5}$$
$$\scriptstyle 1 \; 1$$

But think about how you would add up lengths of time – say, 13 hours 25 minutes and 11 hours 50 minutes.

$$h \; \; m$$
$$13 \; 25$$
$$\underline{11 \; 50}$$
$$\underline{25 \; 15}$$
$$\scriptstyle 1$$

Whether you realized it or not, you will have worked in base 60 to find the answer, because there are 60 minutes in an hour. Furthermore, if you were to express your total in days, you would switch to base 24, and come up with 1 day, 1 hour, 15 minutes.

ECHOES OF ANCIENT BABYLON

Base 10 requires 10 different symbols (including the all-important zero; see page 118) to convey each of the number values: 0, 1, 2, 3, 4, 5, 6, 7, 8, 9. The ancient Babylonians worked in a staggering base 60. They allocated 1 to 10 unique symbols, but then the symbol for 11 was a "10" symbol next to a "1" symbol. The largest digit, 59, was made up of 50 (five "10" symbols together) and 9 (nine "1" symbols). Quite logical, and with the beginnings of our base 10 system built into it.

Why choose such a large base? The reason is almost certainly that 60 has a large number of factors (1, 2, 3, 4, 5, 6, 10, 12, 15, 20 and 30 all divide into 60 exactly), and so it was easy to divide things up.

This ancient but enduring base helps to explain why clocks are built as circles and divided into 60ths, and in geometry and calculus, angles are usually measured in degrees, up to 360° (60 × 6).

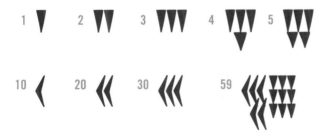

WORKING IN DIFFERENT BASES

Bases may use different numbers of symbols, but they all work in the same way. They may be laid out in columns, or "places", with units (1s) on the right, then the base number, and the value of each place to the left being multiplied by the base number to define the next. For base 10, the places are units (1s), tens (1×10), hundreds (10×10), and so on. For base 3, they are **1**, **3** (1×3), **9** (3×3), **27** (9×3), **81** (27×3) etc. What are the first four place values for base 5?

So how would 176, say, be reworked in base 3? Start with the largest base 3 value below 176 – ie 81. To find the value of each lower place, ask: how many times does this number fit into what remains?

How many 81s in 176?	How many 27s in 14?	How many 9s in 14?	How many 3s in 5?	How many 1s in 2?
2	0	1	1	2
$81 \times 2 = 162$; this leaves 14 ($176 - 162$)	27 is too large to fit into 14	This leaves 5 ($14 - 9$)	This leaves 2 ($5 - 3$)	

So 176 expressed in base 3 is 20112.

THE BINARY SYSTEM *(solutions on page 134)*

Base 2, or the binary system, has a particular importance in science. As only the digits 0 and 1 are used, they can be indicated by "on" or "off", electrical impulse or no impulse, and this is the basis of all computer "brains". The headings are 1, 2, 4, 8, etc. Can you work out:
- how the number 5 would be written in binary?
- how the binary number 110 would be written in base 10?

INTERLUDE
FAMOUS MATHEMATICIANS

Many brilliant mathematicians have helped to further our knowledge by their discoveries. Some are rightly household names, such as Isaac Newton, Archimedes and Albert Einstein, but many more have somehow escaped "superstar status", despite their immense achievements. Here are some of the lesser-known people who have shaped not only mathematics, but the very way we think.

Marie-Sophie Germain (1776–1831) Although a hugely talented natural mathematician living in the 18th and 19th centuries, Germain was discouraged from pursuing a mathematical career because of her gender. Despite a lack of formal training, she managed to study in secret and struck up a long-lasting correspondence and later friendship with C.F. Gauss (see page 24), who praised her work on number theory and Fermat's Last Theorem very highly. For much of her career, she pretended to be a man when corresponding with other mathematicians for fear of not being taken seriously.

Srinivasa Ramanujan (1887–1920) Little known in the West, at least outside academic circles, the Indian Ramanujan was a highly original thinker and largely self-taught. The story is told of his visit to Oxford to meet the mathematician G.H. Hardy, when he arrived in a taxi whose number was 1729. When Hardy remarked that this was a fairly dull number, Ramanujan corrected him, pointing out that 1729 was (of course) the smallest number that could be expressed as the sum of fourth powers in two different ways. But you probably spotted that!

David Blackwell (1919–) Born in the US, Blackwell gained a PhD in mathematics at the incredible age of 22. He is perhaps the most accomplished African American mathematician, and if he had been born 50 years later, he may well have achieved the sort of widespread fame his ability deserved. His forte was statistics, co-writing *Games and Statistical Decisions* in 1954. In 1965 Blackwell became the first African American in the National Academy of Sciences.

William Rowan Hamilton (1805–65) Ireland's most celebrated and brilliant mathematician, Hamilton devised the idea of quaternions, which allowed -1 to have not one but several square roots (see page 131). This discovery had hugely beneficial implications for mathematicians and physicists worldwide. He has never been given the recognition he deserves, partly because his idea is so complex that it is difficult for anyone but mathematicians to comprehend!

Al-Khwarizmi (*c*.780–*c*.840) This outstanding scholar from Persia made significant contributions to geography and astronomy, but is mostly remembered for his ground-breaking work on algebra. Until he started to use letters to generalize numbers, the world had relied heavily on the Greeks' limited geometry-based mathematics. The word "algorithm" is derived from the Latin form of Al-Khwarizmi's name. He also introduced the Arabic numeral system to the West.

John Napier (1550–1617) A wealthy Scottish landowner and mathematician, Napier is wrongly credited with inventing logarithms, which have, in turn, led to some major discoveries in mathematics. He did, however, break new ground when he published *A Description of the Admirable Table of Logarithms* in 1614, which brought logarithms to much wider public attention. Napier was hugely popular, and possibly the closest thing that 16th- and 17th-century Scotland had to a scientific superstar. Many people first come across his work in association with "Napier's rods" or "Napier's bones", a set of sticks that allow complex multiplication to be done very rapidly.

Hypatia of Alexandria (*c*.370–415) Generally acknowledged as the first ever female mathematician, Hypatia was considered sufficiently noteworthy to be painted by Raphael. Religious politics obscure much of the detail of her life, but we do know that she was a mathematics teacher and philosophy lecturer in 4th-century Egypt and that she edited and wrote mathematical texts and commentaries. She was probably put to death by jealous religious zealots who objected to her status, since she enjoyed the favour and respect of men and women alike, owing to her educated mind and graceful manner.

Brahmagupta (598–670) It's hard enough to write a mathematics textbook, but to write one in verse is even more difficult. Yet this is what Brahmagupta did in 628 with his *Brahmasphutasiddhanta*. This is a significant achievement for two reasons. It was the first textbook to refer to zero as a number in its own right, and it was the first book to publish the formula for solving quadratic equations.

Ada Lovelace (1815–52) Daughter of Lord Byron, she was a pioneer of computing. She is quoted as saying, "In almost every computation a great variety of arrangements for the succession of the processes is possible ... one essential object is to choose that arrangement which shall tend to reduce to a minimum the time necessary for completing the calculation." A fitting summary for this book!

EVERYDAY MATHEMATICS

Numbers and calculations fill our daily lives, whether it's just counting out change for a purchase, working out a detailed household budget, recording how much your children have grown, or measuring dress fabric for a bevy of bridesmaids. As you become more alert to numbers, you may also find your interest piqued by the way in which mathematics is integral to so many aspects of life, from the stock market to the satnav system in your car.

From time to time, we are confronted with figures that require greater thought: Is this special offer as good as it says? How can I work out which is the best loan deal on offer? What do various interest rates mean in actual monetary terms? To play ostrich and put your head in the sand rather than get to grips with the reality behind the figures can mean missing out on good opportunities or, worse, being taken for a ride.

MARKET FORCES

Some of our most regular encounters with everyday mathematics are when we go shopping. We can't have total control over how much we pay, but it makes sense to be aware of some of the influences on prices and not be duped by deceptively "good" deals.

ECONOMIES OF SCALE

Why is it that the bigger the shop, the lower they seem to be able to price their goods? It's largely to do with economies of scale.

Suppose I wish to import a car from Japan to sell on to a customer. Every time I order cars, I must pay the factory cost (let's say $6,000 per car), the shipping cost ($1,000), the cost of getting a licence to import cars ($1,000) and finally the rent of a storage facility ($1,000). For a single car, this means that I must charge the customer $9,000 (the unit cost) for the car, just to break even. If, however, I have four customers, they can share the cost of the shipping, licence and storage. This means that the total cost to me is now $27,000, but my unit cost — ie my total cost divided by the number of units (4) — drops to $6,750 — a reduction of $2,250 or 39%.

It's this sharing of costs, when done on a massive scale, that makes it possible for big-scale buyers so often to undercut small-scale buyers. It may not be financially viable to ship, for example, 1,000 dolls halfway round the globe, but if you're dealing in millions, then you can make economic sense of importing from countries that can provide large quantities of goods at low unit cost, even taking into account the shipping costs.

Dealing in big numbers also gives you "buying power". The big supermarkets now purchase 80% of all food grown commercially, and farmers have often been prepared to accept a lower price per unit for their produce because of the guaranteed mass sale.

Economies of scale exist with regard to single products as well. Stores can offer large packs of soap powder or tortilla chips at a better price per weight than small packs, partly because there is a lower ratio of packaging to content.

IS IT A BARGAIN?

It's good to have your mathematical wits about you when out shopping. If money is tight, you need to be sure that you're getting the best deal. Stores have always sought to catch our eye with "unmissable" deals to tempt us to part with our cash. But are these offers as good as they seem? Not always. Exercising the mental arithmetic skills you mastered in Chapter 2 will allow you to tell the difference between the real deal and the sham.

An unattractive but increasingly prevalent acronym in marketing-speak is BOGOF, or "Buy One Get One Free". This is straightforward and (as long as you actually want both of the items on offer) a good deal: "2 for the price of 1".

Another popular offer is "3 for the price of 2". Some people are confused into thinking that this is the same as BOGOF – after all, you appear to be getting one free item in both deals. But do some mental arithmetic before leaping on such an offer. You are, hopefully, going to save one-third of the cost of each item. But if the offer extends to a range rather than identical items, the shop is probably only offering the cheapest one free, and if the unit price is low in the first price, just how much of a bargain is it? And did you want that third item? Bear in mind percentage change, too (see page 54), to help you keep a sense of proportion.

These are not necessarily bad deals, but clearly you need to be wise when shopping for so-called bargains.

SAUSAGES AND MASH *(solutions on page 134)*

Your children have friends coming for tea, and you need 12 sausages and 1kg of potatoes. There are a lot of special offers around:

SHOP	SAUSAGE PRICES	POTATO PRICES
Al's	39p each	26p per 100g bag
Bumper Bargains	53p each (but 4 for the price of 3 today)	We only sell 200g bags for 49p each.
CutYourCosts Corner Mart	45p each (but 10% off today)	We only sell 500g bags for £2.40 each, but today is BOGOF day

You'll need to find the answers to the following questions:
- Where should you get your sausages?
- What about the potatoes?
- Where should you go if you only have time to visit one shop?

BARCODES

You probably con't give them a second glance. But the arithmetic behind those stripes and digits is a fascinating example of the hidden information to be found in the world of numbers.

* * * * *

Barcodes follow a distinct pattern, one of the most common of which is the 12-digit Universal Product Code. A can of ordinary co a needs a different code from a diet cola can, which in turn needs a different code from a six-pack, and so on. Every type of product for sale has a unique barcode. The largest possible 12-digit number is 999,999,999,999, one less than a trillion (one million million, or one thousand billion).

ISBN 978-1-84483-832-5

Each digit is coded by four stripes (white/ black/white/black) whose thickness determines their value ——

Check-sum digit ——

First six digits signify country of origin and manufacturer

Last six signify product itself, and are assigned by manufacturer

The ingenious arithmetic lies in a 13th digit (usually at the start) called a check-sum. For computers to identify items, they must be able to check whether the code has been read accurately. You can do the same:

- First add up all the digits in positions 1, 3, 5, 7, 9 and 11, then multiply the total by 3.
- Next, add up all the digits in positions 2, 4, 6, 8, 10 and 12, and add this total to the result of the multiplication.
- Take the last digit of the resulting total and subtract it from 10. This should equal the check-sum digit.

THE VALUE OF ESTIMATING

Sometimes we can feel overwhelmed by the detail of a mathematical calculation and overlook the fact that all we may actually need is an appropriate approximation. However, once we've realized this, we can then simplify the problem and find the way to an answer.

LOOKING BEYOND THE NUMBERS

When faced with a calculation to do on the spot – perhaps working out how much paint or fabric to buy, or how much to tip a waiter – lots of people feel a sort of number-blindness coming on: how can I possibly work out a percentage of *that*, or where do I even *begin* to turn measurements of walls into pots of paint? Rather than feeling overwhelmed by the minutiae, it's good to develop the habit of looking beyond "all those numbers" and consider calmly what they are actually worth.

As a first step, ask yourself: do I need to know *exactly*, or do I need to know "*around* how much"? Often, the answer is the latter.

Suppose a tip of 12.5% is suggested for a restaurant bill of $68.54. At first sight this seems daunting, but look at the figures with "estimating eyes". Given that the bill is roughly $70, you can say that 10% will be $7, 15% would be $10.50 (half as much again), so 12.5% would be halfway between those, or around $8.50. (The exact answer is, in fact, $8.57, so the quick estimate you have arrived at is certainly more than accurate enough for your purpose.)

Practise this sort of estimating, checking on a calculator when you have a chance, to see how close you are. Practice will enable you to develop a feel for how much to approximate, and will make you alert to what figures actually represent. Get into the habit of rounding numbers up or down to something meaningful and easy to grasp.

UP OR DOWN?

Convention dictates that numbers falling exactly between two markers should be rounded up (so 385 to the nearest ten would be 390 rather than 380) but which way you choose to go, and how close you work to an exact number, depends also on the context: 4,816 would usually be rounded down to 4,800 as it is closer than 4,900, but if it is important to err on the side of generosity (for example, in estimating supplies or allocating accommodation), then 4,850 or 4,900 would be a more sensible working figure.

HOW ROUGH IS "ROUGHLY"?

What makes an approximation acceptable? The key lies in what are known as "orders of magnitude". Orders of magnitude refer to the size of a number. The larger a number, the higher its acceptable margin of error becomes. If a country has a population of 59,470,877 people, for example, then official statistics might record it as 59,000,000. This is actually off by over 450,000, but because this is quite a small fraction of the original it is considered to be an acceptable margin. But why not go to 60,000,000? No reason at all; it all depends on the degree of accuracy required. So 4,820, 4,800 and 5,000 are all good approximations of 4,816, depending on the degree of accuracy required.

A NEW LICK OF PAINT

Here's an example of how estimating – and a little lateral thinking – can bypass much laborious and unnecessary calculating.

The long-winded way to work out how much paint you would need to redecorate a room would be to multiply the height of the walls by the length of the room's perimeter to find the overall area, work out the areas of all the doors and windows, add them up and subtract them from your overall figure, and then divide the answer by your paint's covering power.

A neat trick that is mathematically close enough is to use a door as your estimating guide. The area of a typical interior door is roughly 20 square feet or 2 square metres. These aren't exact conversions, remember, but approximations that give easy numbers to work with. Paint-can labels tell you approximately what coverage you can expect; this typically might be about 100 sq ft per quart or 12 sq m per litre. So 1 quart would paint an area equivalent to five doors (and 1 litre the equivalent of six doors). Now estimate, either by eye or with outstretched arms, how many times your door would fit round the room, and this will give a very good approximation of how much paint you need. (Disregarding window and door spaces usually compensates neatly for the fact that the walls are higher than the door.)

A clever alternative is to measure round the edge of the room in feet and divide by 5 to give you the number of quarts you need (or in metres and divide by 6 for litres). Can you see why this works?

Remember to round up rather than down – oh, and don't forget to allow for a second coat!

SAVINGS AND LOANS

Have you ever been at a loss to work out how much a loan is going to cost you, or how much you'll earn on your savings? The interest paid on one and accrued by the other are the two sides of a financial balancing act, and both are ruled by basic mathematics.

"INTEREST"ING MATHEMATICS

A lender agrees to give a sum to a borrower, who agrees to pay it back in a pre-arranged number of parts – each part being a fraction of the amount borrowed, plus an interest payment. The lender must always quote the APR, the annual percentage rate, by which the exact amount of loan repayments is calculated. If you borrow $100 at an APR of 18%, for example, then interest accrues steadily. If this were simple interest, after one year you would owe $118 if you had not repaid any of the money.

However, this easy-to-understand picture is complicated by two factors. If you regularly repay part of the amount you owe (known as the "principal"), the total you owe goes down, and therefore the amount of interest you owe decreases as well.

The second complicating factor is compound interest. Imagine a savings account that advertises an attractive savings rate of 12% per annum: if you invest $100, how much money would you have by the end of one year? Simple interest would give you $112, but with compound interest (by which almost all savings and loan rates are calculated) you will actually do slightly better.

In compounding, the annual rate is divided by 12, to give you a monthly rate – in the case of your fictitious account, 1%. Adding this earned interest to your initial $100 each month means that, month by month, the money that is earning interest is also growing, as your interest also earns interest.

The following table shows what happens. As you can see, the final amount is somewhat higher than the $112 that you might have gained from simple interest.

MONTH	DEPOSIT	INTEREST	BALANCE
1 Jan	$100.00	$1.00	$101.00
end Jan	$101.00	$1.01	$102.01
end Feb	$102.01	$1.02	$103.03
end Mar	$103.03	$1.03	$104.06
end Apr	$104.06	$1.04	$105.10
end May	$105.10	$1.05	$106.15
end June	$106.15	$1.06	$107.21
end July	$107.21	$1.07	$108.29
end Aug	$108.29	$1.08	$109.37
end Sept	$109.37	$1.09	$110.46
end Oct	$110.46	$1.10	$111.57
end Nov	$111.57	$1.12	$112.68
end Dec	$112.68	$1.13	$113.81

Working out interest *(solutions on page 135)*

There's a simple formula to calculate the final figure: $F = P \times (1 + i)^n$. F is the final figure, P is the principal, i is the annual interest rate, and n is the number of years. The raised n shows that the sum of the figures in the brackets has to be multiplied by itself; so $(1 + i)^3$ would be $(1 + i) \times (1 + i) \times (1 + i)$. Try using the formula to work out which of these investments will yield the larger return.

A €400 invested with compound interest of 5% per annum for 3 years. (Hint: $1 + i = 1.05$)

B €380 invested with compound interest of 4% for 5 years.

SHOP TILL YOU DROP?

Whether you're purchasing a new washing machine, a sofa or a car, you may notice that the salesperson will often be eager to help you "spread the cost of the repayments". Yet at first glance, this makes no sense – surely the store would be better off getting all the money straight away? The answer lies in the commission offered to stores to sell finance packages. Although initially these loans may seem more attractive than paying a huge amount at once, you need to be skilful at deciding whether or not they are the right option for you.

Other than truly interest-free loans, which are always more cost-effective than paying the whole amount up front, you need to be able to calculate how much you'll be paying for the loan in total. To do this, find out the rate of interest (see APR, page 91) and work out what the interest will amount to over and above the cost of the item itself.

A GOOD DEAL –
FASTER

A salesperson can show you figures that seem oh-so-affordable, but how do you decide whether it's really a good deal? Quickly figuring out the actual cost, even approximately, could save your valuable cash.

* * * * *

When you're quoted a monthly repayment, simply go through the following steps.

- First, work out what the APR (see page 91) would be over a year. The easy way to calculate this in your head is: just add a 0 to the monthly figure, for 10 months, then add twice the monthly figure, to give you 12 months.
- A very high proportion of loan repayments are quoted in 36, 48 or 60 months; these are 3, 4 and 5 years respectively, so simply multiply by 3, 4 or 5 to find out the total cost of your loan.
- Subtracting the original cost of the goods will show you how much you're handing over for the privilege of not paying it all up front.

You'll be surprised how much this can help, and more often than not you'll beat the salesperson tapping away on her calculator.

* * *

Don't forget that a small variation in the monthly repayment can make a big difference to the total. Recently, as I was about to sign a loan agreement in a car showroom, I noticed an error in the monthly repayment figure. Although the amount was under £10, I would have ended up paying around £300 more than initially agreed over the course of the loan. Can you estimate how many months the loan was to run?

THE MONEY MARKET

FTSE, Dow-Jones, NASDAQ – phrases we have all heard, but what is the mathematics that drives them? We read about huge wins and losses on the stock market, but few of us understand exactly what is involved. For example, did you realize that in the vast majority of deals, nothing more than promises are actually traded?

THE TRADERS

There is no single entity that can be described as "the market" – it is nothing more than buyers and sellers. Surprisingly, there are far more traders who work for themselves at home than there are traders in banks and investment houses, and despite the mystique surrounding trading, the mathematics required is actually pretty straightforward.

THE MATHEMATICS OF TRADING

Stock prices are always quoted in "points", each of which is worth 1¢ in the USA or 1p in the UK. Traders buy and sell rising stocks, hoping to sell them on before they fall in value. They first calculate what is known as a reward:risk ratio, by noting the most recent high and low values of the stock that they are considering. These are respectively known as the "line of resistance" and the "line of support". When a stock's price is rising, the reward is the difference between the recent high and the current price, and the risk is the difference between the current price and the recent low.

Suppose the figures are 255 points (recent high), 210 points (recent low) and 225 (current price). The potential reward is 30 (255 − 225), but the potential loss is 15 (225 − 210), so the reward:risk ratio is 30:15, or 2:1. Cautious traders will not invest unless the ratio is at least 3:1, so this example would probably not attract much investment.

THE MATHEMATICS OF PROFIT

Next the decision needs to be made as to how much to invest. Each trade should expose only 1% of the investor's capital, so that if it fails (and on average, half will!) it is not a cataclysmic disaster.

How is it possible to make money if half your trades fail? Remember the reward:risk ratio? This means that, if you stick to a ratio of at least 3:1, each successful trade will return at least three times the initial investment. Suppose you make 10 trades, of which seven fail and three succeed. If each trade was for 1% of your net worth, you have lost 7%, but the other three trades each return 3%; a total of 9% and hence a net profit of 2% of your net worth.

Of course, this should not be seen as advice, as nobody is ever guaranteed to make money, but next time you look at the money markets you may have more insight into what is happening.

"Money is like manure, of very little use except it be spread."

Francis Bacon (1561–1626)

NAVIGATION

Since human beings first started to travel around the globe,
there has always been a need to fix our current position
and to know which way to head. Navigation, whether by the
sun and stars or by means of artificial satellites, is based
on mathematical calculation.

TRADITIONAL METHODS

Early sailors and other travellers would use known fixed points, such
as the pole star or a landmark, to calculate their approximate position.
Bearings (see page 99) were invented to give some accuracy to these
directions, using a process known as triangulation.

Out of the sight of land, the only constantly visible objects were
the sun and the stars, and of course these kept changing their
positions throughout the day and night. Knowing that the sun was at
its highest at midday allowed sailors to calculate their latitude (how
far north and south they were). However, not knowing their longitude
(their precise east–west position) was a shortcoming that severely
hampered trade and exploration across the oceans until the mid-18th
century (see box overleaf).

Once sailors reliably knew what the time was in Greenwich,
they could work out the local time using celestial measurements. By
working out the difference between the two times and then multiplying
this figure by 15 (because every 15° degrees east or west represents
an hour's time difference), they could calculate their precise longitude.

The key to calculating longitude at sea rested on knowing the exact time. However, pendulum clocks, the standard design used in the 17th and 18th centuries, could not possibly remain accurate while on board a ship that was swaying around on the waves. In addition, weather conditions such as changes in temperature could render clocks less accurate. In 1714, the British government offered a substantial reward to anyone who could solve the problem, which had puzzled even such great men as Galileo and Newton. The solution was finally provided in 1755 by John Harrison, a clockmaker. His fourth attempt, a pocket watch known later as H4, kept accurate time even under the rolling conditions of a ship because it did not rely on pendulums, and included a mechanism to compensate for fluctuating temperatures.

GLOBAL POSITIONING SYSTEMS

Although our measurement methods have become infinitely more exact and high-tech than those of navigators in ancient times, the mathematics for plotting a location anywhere on the Earth remains remarkably similar.

Global positioning systems (GPS) are little more than sophisticated triangulation systems, which use your position in relation to that of several known satellites orbiting the Earth in order to pinpoint your location (using more fixed points increases the accuracy). Unlike Harrison's pocket watch (see box above), the microchips inside GPS systems can do this several times a second. The systems have built-in maps rather than simply giving degrees of longitude and latitude, but the principle is still the same.

GETTING YOUR BEARINGS

Any position on Earth can be defined by its latitude and longitude. In addition, the direction in which you face or move is defined by a bearing, which is measured in degrees, like angles within a circle.

* * * * *

Imagine that you're standing in the centre of a giant circle on the ground, facing north (or 12 o'clock). This direction is known as bearing 000 degrees (bearings always have three digits). If you turn 90 degrees to your right, for example, this is bearing 090°.

* * *

You can take bearings from known landmarks to locate yourself on a map. Suppose you see a church to the west of you and a lake to the north-east. If the church is west of you, you must be east of the church (bearing 090°), so draw a line on the map running eastward from the church. Using similar reasoning, draw a line south-west from the lake (225°). The point where the two lines intersect will show your location.

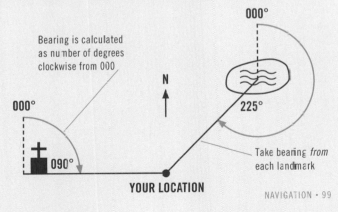

Bearing is calculated as number of degrees clockwise from 000

000°

N

000°

225°

090°

Take bearing *from* each landmark

YOUR LOCATION

CHAPTER 5

NUMBERS: CAN YOU REALLY COUNT ON THEM?

At an everyday level, numbers are concrete and reliable, but some of the uses to which they are put can make them seem slippery and untrustworthy. The cry "Lies, damn lies, and statistics!" was a heartfelt one. However, this isn't the fault of numbers, but of interpretation and context — for example, "50 per cent" is quite meaningless unless we know 50 per cent of what. Misunderstanding of statistics or fear of tackling numbers head-on can lead us to be seduced into believing almost anything. This chapter aims to pull some of the wool from your eyes and show you how sometimes mathematical fact can be at odds with your instinctive assumptions.

STATISTICS: SHOULD WE TRUST THEM?

People tend to treat statistics in one of two ways: either they accept the findings unquestioningly, or they cynically reject them. Which is the healthier option? The former attitude is based on ignorance, and the latter arguably on fear – but our best bet is a questioning mentality, in order to find out what the statistics are actually telling us.

WHAT STATISTICS ACTUALLY TELL YOU

Statistics often involve percentages, and particularly percentage change, which, as we have seen in Chapter 3 (page 54), can present a confusing picture. Figures are frequently quoted with a lack of rigour, based on laziness or misunderstanding. You might have read in the press, for example, that one extra alcoholic drink per day increases a woman's risk of breast cancer by 6%. Such a threat might be sufficient to stop many women drinking altogether, yet a bit of digging reveals that women are extremely unlikely to develop breast cancer from this extra alcoholic drink. In fact, 9% of women will get breast cancer by the age of 80, and 6% of 9% is just under 0.5% ($0.06 \times 0.09 = 0.0045$ or 0.45%).

Raw percentages, even when accurate, are tricky to visualize, so it's a good idea to translate them into terms of people per hundred, or per thousand. In the above example, 0.5% is easier to grasp as 1 in 200 women. So a clearer (and much less scaremongering) statement of the same statistic would be: "Typically, 18 out of 200 (9 out of 100) women would develop breast cancer by the age of 80. If they all drank one extra alcoholic drink per day, that figure would rise to 19 out of 200.' Always be alert to what percentages actually mean, and ask yourself: X per cent *of what*?

CAUSATION OR CASUAL CORRELATION?

Another common pitfall is the assumption that if A increases, followed by a similar increase in B, then A must have caused B. As there are many examples in which A and B *are* linked, it's all too easy to fall into the trap of believing that this is always so. In an amusing example from Scandinavia, researchers found that the more storks that nested on your family's roof, the more children you had. There was no proof that the storks caused the children or vice versa. A bit of thought suggests that the more children you have, the bigger your house is likely to be, with a larger roof to attract more storks! It's a little like saying that global temperatures are directly related to the number of Elvis impersonators, since both have risen significantly in recent decades. It's all too easy to misuse the numbers presented to us.

DATA COLLECTION METHODS

Statisticians often work with surprisingly small samples. Although this may be counter-intuitive, it's generally true that the larger the population being sampled, the smaller the percentage needed for an accurate result. For example, a sample size of 500 people would provide at least as valid a result for a population of 5 million people as for 1 million, provided that the sample is as random as possible.

TRUE RANDOMNESS

A common concept of randomness is of a wide distribution, or a selection that shows no favouring of one area or type over another. A random selection of US state representatives should not exclude anyone west of the Rockies; a random collection of clothing should not be all socks. Lottery players may equate picking numbers at random with choosing a wide variety rather than just a cluster of adjacent numbers. Yet randomness is very different from an even spread. Imagine scattering rice over a square-tiled floor. There will be some tiles with lots of rice and some with very little. This misperception of the nature of randomness explains the panics caused by, for example, apparent "cancer clusters": places where the incidence of cancer is much higher than average. In fact, reassuringly, most clusters are simply random mathematical events, rather than reflecting genuine concentrations of danger.

However, even statisticians don't consider statistics an exact science, and figures are published with what is known as a margin of error – an indicator of how reliable the statistics are. A larger sample may enable statisticians to declare a higher confidence level in their results (ie how confident they are that the whole population's results would mirror those of the sample) and reduce the margin of error.

MARGINS OF ERROR

If, for example, a poll says that there is a 4% margin of error, that means that every 25 times you repeat the survey, you'd get on average one totally spurious response – there's a 4% chance that it might just be the one you are looking at!

Margins of error are important when defining what results mean. Suppose a politician's popularity rating is given as 46% in a poll and 48% when the poll is repeated a week later. This doesn't necessarily mean that her rating has risen. If the margin of error in both polls is, say, 3%, then the first result should actually be read as 'between 43% and 49%"; the second result also falls within this range.

SO CAN WE EVER TRUST STATISTICS?

Absolutely, provided that we never take them at face value, and always ask the right questions. When you see a percentage, always ask *what* it is a percentage of, what the measurement criteria were, what the sample size and margin of error might be, and whether there were any other influencing factors. This way, you'll be able confidently to assess the validity of what you read.

IT'S ALL A MATTER OF SCALE

Numbers, especially comparisons such as sales figures or how budgets have been allocated and used, are often presented in graphic form. Pie charts, graphs and histograms can all be useful ways of making such figures more understandable, but they also present wonderful opportunities for misrepresentation.

READING THE CHARTS

When faced with the prospect of delivering worse than expected figures, it is very tempting for individuals, companies and even governments to put as positive a "spin" as possible on the actual numbers. Consider the sales diagram below, created by Mr Widget. It seems that his manufacturing company has performed magnificently over the past two years – at first glance, it appears that sales have more than doubled.

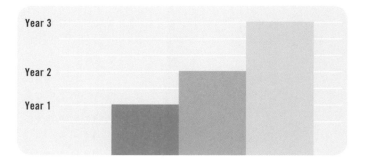

However, what is conveniently missing are the all-important labels on the vertical axis. Instead of starting at zero (the honest thing to do), Mr W has only shown the top of his sales graph. If each horizontal line on the graph were to represent, say, 100 sales, and his sales in Year 1 amounted to 5,000 items, the sales in Year 3 (5,500 items) would actually be an increase of only 10% over those in Year 1.

Graphic representations are easy to manipulate, so always check the base lines and the actual information given, not the just the effect created. Simple graphs can be manipulated to exaggerate or minimize relative changes. Similarly, pie charts always give an impression of showing a whole, but may just represent part of the story.

Look at these two graphs plotting Annabel's and Barbara's New Year attempts at losing weight. Who gives the impression of having done better? Certainly Annabel's graph is steeper, suggesting more rapid progress, but look closely at the horizontal axes: how different is the picture when you look beyond appearances to the actual figures? How much weight have they both lost by 15 January, for example?

CHANCE AND PROBABILITY

Probability determines not only the odds on a horse race or the chance of your winning the lottery, but also mundane practicalities such as how much insurance you pay or how high a risk you run of suffering certain accidents or diseases.

THE PROBABILITY CONTINUUM

Probability is one of the most misunderstood areas in all of mathematics: at best its laws are less than obvious, and at worst they are highly counter-intuitive.

Imagine a horizontal line with 0 on the left and 1 on the right. We can map all future events somewhere along this line according to the following principle. If something cannot possibly happen, it has a probability of 0, or 0%; anything that is certain to happen has a probability of 1, or 100%. All other events belong somewhere between the two. Something that has an even chance of taking place (for example the tossing of a coin and getting "heads") has a probability of exactly 0.5, or 50%. The weather forecaster who predicts a 50% chance of rain is actually saying that it is equally likely to rain as not

Impossible	Very unlikely	Unlikely	Even chance	Likely	Very likely	Certain
0			0.5			1
0%			50%			100%

rain – not terribly useful! And a common error that people fall into, perhaps because 1 in 10 is equivalent to 10%, is to think that, for example, 1 in 20 is equivalent to 20%; of course, it is actually 5% (as 1 in 20 means 5 in 100).

HOW PROBABILITY IS CALCULATED

The probability that an event will happen is calculated by dividing the number of ways that the event could happen by the total number of possible outcomes. This sounds more complicated than it really is. Suppose you want to work out the probability that throwing a die will score *more* than 4.

- How many ways it could happen: 2
  (the die could show a 5 or a 6).
- How many outcomes are possible in total: 6
  (it could show 1, 2, 3, 4. 5 or 6).

So the probability of scoring more than 4 is $2 \div 6$, or $^2/_6$. This is the same as $^1/_3$, so we can assume that we will score a 5 or a 6 approximately one in three times.

Why approximately? There's an important difference between *theoretical probability* and *measured outcomes*. The more times you carry out an action, the closer your results will come to the theoretical probability. If you were to throw your die an infinite number of times (clearly impossible in reality), you would find that a 5 or 6 came up exactly one-third of the time – life mirroring theory precisely. But if you threw it a mere hundred times, or even a thousand times, the best you could say would be that the *chances are* that a 5 or 6 would come up about one-third of the time. It's this element of chance on which gambling (including trading on the stock market) is based. This is why past records, whether it's a horse's racing form or the highs and lows of a company's shares, are important in trying to predict as accurately as possible future performances or outcome, but are no *guarantee* of what will happen in the future.

The laws of probability affect our everyday lives more than you might expect. They dictate, for instance, the cost of insurance. Actuaries work out how likely you are to have a car accident or have your house broken into by calculating the chances of certain events occurring – the same rule as forecasting the chances of a die falling on a particular number, but with more variables. Your car insurance premium will be based on criteria such as age, type of vehicle and where you live, as well as balancing the chance of the insurance company losing business if the quote is too high and going out of business if it is too low to cover losses through claims.

THE SHARED BIRTHDAY CONUNDRUM *(solutions on pages 135–6)*

The chances of certain events are often very different from your intuitive guess. One of the most famous probability conundrums has become known as the Monty Hall problem (see page 112), but here is a more straightforward question of chance: how many people must be in a room before it is more likely than not (ie there is a probability of greater than 50%) that at least two of them share the same birthday? 100? 500? Incredibly, the answer is just 23. How can this be?

- The probability that you and I share a birthday is 1 out of 365. The probability that my birthday falls on a day that is *not* your birthday is therefore 364/365, or about 99.7%.
- The probability that the birthday of a third person falls on one of the remaining 363 days is 363/365.
- A fourth person has only a 362/365 (about 99.2%) chance of avoiding all our birthdays.
- Following this through, the 23rd person has a 346/365 (about 93.7%) chance of avoiding all the other birthdays.

To calculate the probability of two things happening together, we need to multiply the probabilities of them happening separately. For example: the probability of tossing a coin and getting "heads" three times in a row is $1/8$ ($1/2 \times 1/2 \times 1/2$), or 12.5%.

If you multiply all the birthday probabilities: 364/365 × 363/365 × 362/365 and so on, by the time you reach 346/365 (the 23rd person), you reach an answer of 49.3%. As this is slightly less than half, you've reached the tipping point at which it becomes slightly more likely than not that two people in the group share a birthday.

The reason why this is so widely misunderstood is that, when faced with this problem, most people misrepresent it in their brain as, "How many people would have to be in the room before it was likely that one of them shared a birthday with *me*?" Not the same question at all. But can you work out the answer?

And if your birthday is on July 1, you share a birthday with me – what are the chances of that?

THE MONTY HALL PROBLEM

This now famous statistical problem is named after an
American game show host, and has confounded some
very sharp brains indeed.

Suppose you are on a game show and must guess which of three doors
conceals the star prize: a car. Behind each of the other two doors is a
goat. You indicate a door. The show's host (who knows where the car is
located) opens a *different* door from the one you've chosen, to show a
goat. You are then offered the chance to stick with your original choice
or to switch to the other closed door. Is it to your advantage to switch?

The intuitive answer is that sticking or switching makes no
difference. After all, there are now two closed doors – one hiding a
goat, and one hiding the car. The odds must surely be 50:50, or equal.
Amazingly, you *should* switch, as you're then *twice as likely* to win.

At the start, you're equally likely to make any of the three choices:

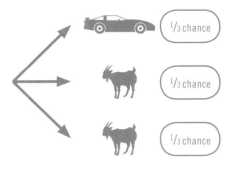

But see what happens at the second stage, depending on whether you switch or stay put:

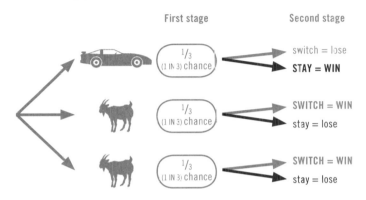

First stage   Second stage

$^1/_3$ (1 IN 3) chance   switch = lose

STAY = WIN

$^1/_3$ (1 IN 3) chance   SWITCH = WIN

stay = lose

$^1/_3$ (1 IN 3) chance   SWITCH = WIN

stay = lose

As you can see, if you switch from a goat you'll win, while if you switch from a car you'll lose. Since you'll select a goat, on average, every two out of three times you try, then two-thirds of the time, switching will result in a win.

You can try this yourself with three playing cards and a friend. Use an ace to represent the car, and two jokers for the goats. Fan the cards in front of you so that only you can see their faces. Ask your friend to point to a card. Reveal one of the jokers she did *not* choose. Ask if she wishes to change her mind and switch from her original card. Reveal her final choice and make a note of whether she wins or loses. You'll find, if you play enough times, that by switching she'll win roughly twice as often as when she chooses not to switch.

WHAT ARE THE ODDS?

Mathematics is at the heart of all gambling odds, but can a better understanding of your chances actually give you an advantage, or is everything ultimately down to luck? Is a lottery just a tax on poor mathematicians?

THE BALANCE OF PROBABILITY

Betting and games of chance are all about the balance of probability. At a roulette table, the biggest payouts are on bets placed on a specific number, but the chances of that number coming up are low (1/36, or 1/38 on some wheels). Opting for a simple odds or evens bet carries a high chance (1/2) of winning, but the payout is low. Bookmakers use the same balancing principle when setting the odds on a horse race, but you can improve your chances of choosing the winner if you know something of the horses' previous results.

While some forms of betting can be influenced by skill and experience, others are purely luck. One popular type of betting, enjoyed by many people who'd never consider themselves gamblers, falls into the latter category: the lottery. Because of the element of luck as well as popularity in the lottery, it's interesting to look at the mathematics.

THE LOTTERY – IS IT JUST A LOTTERY?

All lotteries are broadly similar: you pay to choose a set of numbers, and you win if your numbers match any or all of an official set drawn at random. In the UK Lottery (and in US state lotteries run along very similar lines), a choice of 49 numbers means there are 13,933,816 equally likely combinations. The chance of your combination coming up, therefore, is almost 1 in 14 million – not encouraging odds.

Whatever your intuition may tell you, there's nothing you can do to increase your chances of winning. But you *can* maximize the amount of your prize, should you be a winner. The key lies in the fact that the jackpot is shared among *everyone* who has matched the winning combination of numbers. Imagine finding you've won the jackpot, only to discover that you'll be sharing it with 5,000 other people. So the trick is to choose numbers shared by as few people as possible, so that if you do win, you'll receive a larger fraction of the available pot.

Pick a number, any number ...

People tend to form illogical attachments to certain numbers or groups, such as significant dates or "special" sequences (several thousand every week choose 1, 2, 3, 4, 5, 6). So to reduce the potential number of co-winners, include at least some figures above 31, avoid those forming obvious patterns, and aim for randomness. Remember, though, that selecting numbers that appear evenly spaced around your ticket is not the same as choosing at random; indeed, "choosing randomly" is, mathematically speaking, an oxymoron (see page 104).

"Fortune truly aids those of good judgment."

Euripides (*c.*480–406bc)

THE WONDER OF NUMBERS

We mostly see numbers as a means to an end—workaday soldiers that are there for us to deploy in often humdrum but necessary calculations. But they have a life of their own, too: they relate to each other in surprising ways, and are at the heart of some of the most amazing natural patterns.

Some numbers, or relationships between numbers, also have properties that have fascinated mathematicians from earliest times. The realms of mathematics include not only "real numbers", the digits we all know, but unreal numbers, irrational numbers, and even imaginary numbers. And the rules by which numbers exist may yet provide the key to unlocking secrets of the universe in the future.

ZERO: SOMETHING AND NOTHING

Opinion is divided among philosophers as to whether "nothing" can actually exist, but mathematicians are in no such doubt. Zero is a hugely important number, and its invention heralded great strides forward in many areas of mathematical thinking.

BEFORE ZERO

To appreciate the importance of zero, it's first necessary to recognize how tricky arithmetic was before its introduction. When writing numbers with more than one "place" (see page 75), a 3, say, might mean 3, or 30 or perhaps 300 in our decimal (base 10) system; in other bases, it was more complicated still. To overcome this problem, the ancient Babylonians introduced two slanted marks to indicate where an empty "place" occurred in the middle of a number: ◀◀ Such a symbol is known as a place-holder, and this is still an important function of today's zero.

0 JOINS THE NUMBERS LINE-UP

"Nothing" was seen as merely "an absence of everything", however, until Indian mathematicians (not the Arabs or the Babylonians, in contrast to commonly held misconceptions) realized around the 7th century AD that zero was a number in its own right that could be placed on the number line, and behaved (in some ways but not in others) like a number. It acquired the symbol 0, and this was spread by Arab traders – who are often therefore credited with its invention. In truth, the Indians, Arabs, Babylonians and even the ancient Greeks were all involved with the development of what has become a key mathematical idea.

Inexplicably, the symbol 0 fell out of popular use in antiquity, even among mathematicians, and it was not until the 17th century, almost 1,000 years later, that it was taken up again.

THE STORY OF 0

The word "zero" originated from the Sanskrit *sunya*, meaning "empty" or "nothing". The number gradually acquired its name as it passed along the trade routes of Asia and Europe. The Arabs named it *sifr* (from which we derive the word "cipher", meaning a type of secret code). The Arabic *sifr* entered Latin as *zephirum*, which became *zefiro* in Italian; the medieval Venetians changed and shortened this to make "zero".

HOW HAS ZERO HELPED MATHEMATICS?

Apart from its usefulness as a place-holder, once mathematicians realized that 0 was a real number, it became instrumental in algebraic systems to solve ever more complex equations. Because adding or subtracting 0 doesn't change the value of another number, it became possible for mathematicians to manipulate expressions more easily and resolve hitherto unprovable hypotheses. Also, our understanding of zero has helped to develop our thinking about infinity and limits, and this, in turn, was instrumental in the development of calculus, one of the most powerful new mathematical ideas of the second millennium.

ZERO'S ODD
BEHAVIOUR

It will be obvious that adding or subtracting zero doesn't change anything: 6 + 0 = 6, and 1,546 − 0 = 1,546. But what about multiplying and dividing?

*   *   *   *   *

Multiplying by zero often trips people up, but is also a reasonably easy concept to grasp: $8 \times 0 = 0$, because eight lots of nothing is still nothing. Division, however, poses a more interesting challenge.

Currently, dividing by zero has no meaning. I say "currently" because it's possible that in the future some brilliant mathematician might discover a new way to think about this concept. But for now, ask yourself: how many times does 0 go into 8? Certainly not 0 times, since $0 \times 0 = 0$, but not 8 either, and no other answer makes any greater sense.

If we were allowed to divide by zero, we could prove ridiculous things. For example:

$4 \times 0 = 5 \times 0$

This mathematical statement is true, since both sides of this equation equal 0. But dividing both sides of the equation by 0 would give us $4 = 5$ ... an impossible contradiction.

The mathematicians' solution to the problem is simply to define the process of division by zero as a meaningless and impossible one.

*   *   *

If a weather report helpfully informs you that the temperature is predicted to be twice as cold as today, and today is zero degrees, what will tomorrow's temperature actually be? *(solution on page 136)*

PRIME NUMBERS

All numbers have certain properties. For example, all whole numbers have factors – numbers which divide into them without remainders. Numbers with exactly two factors (themselves and 1) are known as primes, and they have fascinated mathematicians for centuries.

THE SIEVE OF ERATOSTHENES

Even the early Greeks knew about prime numbers, and a Greek mathematician named Eratosthenes (c.276–194BC) discovered a very simple way of working them out.

- Eratosthenes first wrote out the numbers from 1 to 100.
- He crossed out 1 as it only has one factor and is therefore not a prime number.
- He then circled 2 (as his first prime) and crossed out all the multiples of 2.
- The next uncircled number, 3, must be prime, so he circled it and then crossed out all the multiples of 3.
- The next uncircled number, 5, must be prime, so he circled that and then crossed through all the multiples of 5.
- He repeated the process with primes from 7 and upward.

This technique resulted in the table shown overleaf (primes have been highlighted rather than circled). It has come to be known as the sieve of Eratosthenes.

Number 1 is crossed out as it has only one factor – itself – so is not prime

Highlighted numbers are prime – their only factors are themselves and 1

Crossed-out numbers are non-prime, having three or more factors

1	2	3	4	5	6	7	8	9	10
11	12	13	14	15	16	17	18	19	20
21	22	23	24	25	26	27	28	29	30
31	32	33	34	35	36	37	38	39	40
41	42	43	44	45	46	47	48	49	50
51	52	53	54	55	56	57	58	59	60
61	62	63	64	65	66	67	68	69	70
71	72	73	74	75	76	77	78	79	80
81	82	83	84	85	86	87	88	89	90
91	92	93	94	95	96	97	98	99	100

This process satisfactorily identifies the prime numbers below 100. It is also possible to use Eratosthenes' sieve to find higher prime numbers, and with the help of supercomputers mathematicians are constantly searching for ever-higher primes.

A PATTERN FOR PRIMES?

Primes are defined as having no factors other than themselves and 1. What is not so well known is that, apart from 2 and 3, all prime numbers are either one more or one less than a multiple of 6. There is a very neat proof of this: briefly, all numbers other than those one place higher or lower than a multiple of 6 are multiples of either 2 or 3, and therefore by definition cannot be prime.

One of the fascinating and frustrating things about primes is that they don't conform to any discernible pattern. Mathematicians have been intrigued by this fact to such an extent that a $1 million prize is currently on offer to anyone who can solve what is known as the Riemann Hypothesis. This is a possible connection between primes and non-primes, postulated by the German mathematician Bernhard Riemann (1826–66) in 1859. To date this remains perhaps the most popular unsolved mathematical problem in the world. Will it ever be possible to *predict* what the next prime number will be?

PRIME SECURITY

Primes' exclusivity of factors makes them highly inflexible. They can't be broken down into any equal fractions, so they're pretty useless as a basis for currency or any measuring system. However, this very inflexibility has been found to be of great interest in certain commercial applications, most notably credit card and web security.

Security systems often rely on a technology known as "public key cryptography". This has its roots in a simple fact about large prime numbers: they are very straightforward to multiply together but, given the answer, it's almost impossible to find the two original primes. This only works if the primes are large enough, though – primes of 20 digits or longer are commonplace in such technology. Think of this as equivalent to an electronic padlock. You can lock it (by multiplying two primes together), but only someone who has the key – ie knows which two primes have produced this incredibly long number – can unlock it again. This is what keeps electronically transmitted messages secure. Next time you send an e-mail, or make a private phone call or a secure internet purchase, thank primes for the fact that you're well protected.

CODES: A MATHEMATICAL ENIGMA

Ever since people began to use written communications, it became desirable to send messages privately or secretly, and so the art of encryption (translating messages into secret codes) was born. Naturally, mathematics has been instrumental in the setting and cracking of many codes.

MATHEMATICALLY MULTIPLYING PERMUTATIONS

The simplest codes are known as substitution ciphers. These simply involve swapping one letter for another, and are easy to crack. Picture the alphabet written out twice, on two strips of paper curled around to form wheels. Rotating one or both wheels so that identical letters are no longer aligned creates a letter-substitution table. After moving one wheel on by one letter, for example, the picture would look like this:

A	B	C	D	E	F	G	H	I	J	K	L	M	N	O	P	Q	R	S	T	U	V	W	X	Y	Z
B	C	D	E	F	G	H	I	J	K	L	M	N	O	P	Q	R	S	T	U	V	W	X	Y	Z	A

If our secret message to be sent is the word "MOUSE", by replacing each letter in the top row with the letter below it, M O U S E becomes N P V T F. But this code is hardly secure, as there are only 26 possibilities. Even random substitutions made by rearranging the letters on the lower wheel can be cracked by an intelligent decoder.

The code becomes stronger if the letters are not in sequence, and the number of rotations changes, not only for each message but each time a letter is coded. Thus, an A might stand for B in one word but W in the next. This was the method employed by the designers of the Enigma machine (see box below), which used not one but five wheels: part of the encryption involved which three wheels were chosen to fit into the machine. This may sound simple, but mathematically speaking it meant that the wheels could be placed in over a million different permutations, and was the reason why it took the greatest mathematical minds of their generation and a bit of luck to finally crack the code. (In fact, the Enigma contained more complications to further randomize the cipher to several trillion possibilities!)*

Try to crack this more straightforward cipher:

Brx duh d qxpehu zlcdug!

THE ENIGMA MACHINE

It's quite possible that, without the help of their mathematicians, the Western allies might well have lost World War II. The Enigma encryption machines, whose codes were believed to be unbreakable, allowed the Germans to send messages freely between themselves with little or no fear of their messages being read by the allies. So important was breaking the Enigma's codes that Britain's finest mathematical brains, notably Alan Turing (1912–54), were hidden away at a top-secret government facility in Bletchley Park, near London, to work on this. After huge effort, they succeeded in cracking the encryption system and, in the course of doing so, made leaps forward in the technology of computing.

*The exact number of possibilities is $5 \times 4 \times 3 \times 26 \times 26 \times 26$. Can you work out why? *(solutions on page 136)*

UTTER CHAOS!

You may have heard it said that when a butterfly flaps
its wings in South America it can cause a tornado in New
York. This is a good, if somewhat misleading, illustration
of a principle that has come to be known as chaos theory.

WHAT IF?

In simple terms, chaos theory is a branch of mathematics that
explores why random events occur for no obvious reason. At its heart
is the idea that very small changes in a system can cause a ripple of
ever more significant changes that are quite disproportionate to the
original variation.

In the 1960s, Edward Lorenz (1917–2008), a meteorologist, was
working on equations that predicted weather patterns, but discovered
that a tiny difference in his initial data resulted in massive changes

THE PERFECT GOLF SWING?

A simple way to visualize the idea of chaos is to imagine a professional
golfer on the tee. If he hits what appear to be 10 identical shots, in
theory his ball should land in the same spot every time. It does not, of
course, owing to tiny differences such as a puff of wind, a quarter-inch
of extra backswing, or a fraction-of-a-degree turn in his body. Any
of these factors can make a significant difference to the ball's final
position, and of course two or more of them together can make the
outcome even more unpredictable. Nobody can be sure of exactly where
the ball will land each time, thanks to its sensitivity to minute variables.

in the outcome of his experiment. It was he who memorably named this phenomenon the "butterfly effect".

Even Hollywood has picked up on this idea, with thought-provoking films such as *Sliding Doors* (1998) and *The Butterfly Effect* (2004). What if you could go back in time and change one tiny thing — how differently might your life have turned out? What if you never looked across the room and never met your spouse? What if your parents had never met?

Because Lorenz was primarily a meteorologist, he was unable to publish his findings in any mathematical journals, and submitted them instead to a meteorological journal, which meant that his ideas were not discovered by mathematicians until years later. Chaos theory has since become incorporated into predictive modelling in fields such as climate change, economic forecasting and public health planning.

BREAKING THE PATTERN

Lorenz developed equations of how systems behave when subjected to minute unpredictable variations, and these threw up further surprises. He found that the graphs of his equations never seemed to repeat, but instead developed surprising results known as fractal patterns.

The word "fractal", which shares a Latin root with "fracture", refers to the multi-fragmented shapes with which this re atively recent field in mathematics is concerned. Fractals defy the rules of regular Euclidean geometry (circles, triangles and the like), but follow their own fascinating geometry based on potentially infinite repeats at different scales. We will discuss these extraordinary patterns next.

DISCOVERING FRACTALS

Although something was known about the idea of fractals as early as the 17th century, the term "fractal" was first used in 1975 by the French mathematician Benoît Mandelbrot (1924–), to describe shapes which, when examined under a mathematical microscope, appear the same. This is because each time we zoom in, we find the pattern of the object repeated on a smaller scale.

COMPLEX PATTERNS IN NATURE

Have you ever studied a fern? Each frond is made up of many leaflets, each of which is a miniature version of the main frond. Look closer and you'll see that each leaflet is, in turn, made up of yet smaller leaflets, and these, too, are almost identical to the larger frond.

Another well-known example is the snowflake. The arms of snowflakes resemble fractal patterns because at varying degrees of magnification the shape of the arm is replicated, and so these shapes appear identical at any scale.

A surprising feature of fractals is that, while traditional geometrical shapes have a perimeter of a defined length, the perimeter of a fractal such as the Koch snowflake (see opposite) can be infinite. To understand how this arises, imagine how, from space, the island of Iceland might look like a simple ellipse, but as you get closer its outline appears as an ever longer, more convoluted set of headlands, inlets and fissures.

THE KOCH SNOWFLAKE

This fascinating example of fractal geometry is mathematical rather than natural in origin. It is named after the Swedish mathematician Helge von Koch (1870–1924), who first discovered it in 1904.

* * * * *

Starting with an equilateral triangle, remove the central third of each line and replace it with two lines of the same length, to form another equilateral triangle attached to each side of the original. Now repeat with each line on the new shape. And then repeat again. Doing this just twice gives the following progression:

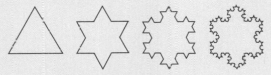

It's possible to continue this pattern forever.

There's a very interesting twist to this particular shape. The area of the first, star-shaped Koch Snowflake is exactly 1.6 times that of the original triangle. Yet however many more times you repeat the pattern of adding new triangles to the edges, the area (calculated using what are known as geometric series, a little beyond the scope of this book) remains exactly 1.6 times that of the original triangle. The logical assumption is that the perimeter can become infinitely long but the area it encloses is unchanging!

A WORLD WITHOUT NUMBERS?

Imagine not knowing how much money to hand over, or what change you should receive, while shopping. Imagine trying to play any sport that involves scoring if you find it hard to grasp that 16 is bigger than 12. Imagine the difficulty of getting to a meeting at the right time, or sticking to the speed limit, if numbers are a foreign language to you.

ACALCULIA AND DYSCALCULIA

For some people, numbers have very little meaning at all. The brains of acalculics ("alcalculia" means "without number") have great difficulty in counting small groups of objects, a skill known as "subitizing", and they struggle with both number sizes and number relationships. They may not be able to do simple sums, and may even be unable to count or to put a list of numbers in size order.

Acalculia is usually brought on by some form of brain injury, most often a stroke. It can also result from genetic impairment, hampering a child's comprehension of numbers. Unless the problem extends to full acalculia, it is called dyscalculia: like the better-known dyslexia, it can vary in severity. "Number blindness" can't always be explained as a genetic trait, but true dyscalculia is a constant frustration, since numbers inhabit every area of our lives. Knowledge about dyscalculia is still emerging, and hopefully in the future we'll understand better why some brains struggle with concepts of number that most grasp almost instinctively.

INTO THE WORLD OF THE UNREAL

We've already met some numbers that are less than concrete – pi (π) and phi (Φ), for example. But in the upper realms of mathematics there exist some very strange numbers indeed … one of which, called "i", cannot even be described as real.

IMAGINARY "i"

When you multiply a number by itself, you obtain a square number: for example: $5 \times 5 = 25$. The inverse of this is the square root, so $\sqrt{25} = 5$. You may also remember from school that multiplying two negative numbers together always results in a positive number. So think about this: if $1 \times 1 = 1$ and $-1 \times -1 = 1$, can $\sqrt{-1}$ exist?

In fact, the answer is yes. For centuries, mathematicians had found $\sqrt{-1}$ appearing in complex equations. Since these equations had a real, concrete answer, $\sqrt{-1}$ had to exist in some form. By the 18th century, $\sqrt{-1}$ had been designated i, and all such numbers (square roots of negatives) became known as imaginary. Acknowledging their existence furthered progress in physics, engineering and electronics. Without i, we'd have no computers, cars, televisions or mobile phones.

"This miracle of analysis, marvel of the world of ideas … we call the imaginary number."

Gottfried Wilhelm Leibniz (1646–1716)

SOLUTIONS

CHAPTER 1
p20: 5 or V or 伍

- The answer is 2: the numbers refer to opposing faces on dice.
- $5{,}550.55 - 500.55 = 5{,}050$; all the others $= 5$.
- In temperature conversion: $5°C = 41°F$ (and $5°F = -15°C$).

pp30–31: Progress?

1   a: 306 (340 − 34)
   b: 27.93. (Begin by multiplying 4×7 to make 28, then subtract
      10 lots of 0.01.)
2   a: 300 (since there are 300 halves in 150)
   b: 1,000
3   $100 \times 100 = 10{,}000$
4   They are equivalent temperatures ($61°F = 16°C$, $82°F = 28°C$).
5   a: 403 (nth term is $4n + 3$)
   b: 49 terms
6   **START 13**   143   144   12   1,186   118.6   237.2
    START 340   170   85   17   68   340. Note that you end up back
    at the beginning, as you've effectively divided by 20 and then
    multiplied by 20, thus "undoing" the first three steps.
7   32145. (This is a Latin square – see page 26.)

CHAPTER 2

p39: Quick calculation chains

START 20   2   4   48   14   42   6   36   3,564

START 7   49   539   100   10   2   8   64   70

p40: Pattern spotting

A   8 (Multiply the first two numbers and subtract 2 to get the third number.)

B   3 (Third number is twice the sum of the first two numbers.)

C   19 (Third number is first number squared plus second number.)

p40: Crossed lines

p41: Kakuro

			16\	17\	14\		23\	24\
	18\18	7	9	2	25\	16\	7	9
28\9	7	9	8	3	1	17\	9	8
17\	8	9	17\31	8	9	8\17	1	7
3\	1	2	20\17	3	1	2	8	6
16\13	9	7	17\30	8	9		24\	16\
31\24	2	8	9	7	5	16\	9	7
8\	7	1	17\	8	9	16\17	8	9
16\	9	7	34\	4	6	9	8	7
11\	8	3		24\	8	7	9	

p61: Figuring with ratios

- The ratio of brown to black bats is 30:20, or 3:2. As 200 is 4 × 50, multiply both parts of the ratio by 4. So 30 × 4 = 120 brown bats.
- First, find out how many parts, or shares, there are in total: 5 + 6 + 7 = 18. 90 ÷ 18 is 5; so each share comprises 5 sweets. Therefore, William gets 25 sweets (5 × 5), Thomas gets 30 (5 × 6) and Rebecca gets 35 (5 × 7).
- To find the proportion of plasma in blood, subtract the proportion of blood cells (45%) from the full amount of blood (100%). 100 − 54 = 55. The ratio of plasma to cells is 55:45; this can be expressed more simply if you divide both sides by 5, to give 11:9.

p77: The binary system

- 101
- 6

CHAPTER 4

p86: Sausages and mash

- Al's offers the best price on sausages: 39p × 12 is £4.68. At Bumper Bargains, you only pay for 9 rather than 12 sausages, but at 53p each this works out at £4.77. At CutYourCosts Corner Mart, you pay 10% less than the full price (which, at 45p × 12, would be £5.40), but even so this works out at £4.86.
- CutYourCost's has the cheapest potatoes – you get your full 1kg for only £2.40. At Al's, you'd be paying for 10 × 100g bags, which

would be £2.60. At Bumper Bargains, you'd have to pay for 5 × 200g bags, or £2.45.

- Interestingly, you'll get the lowest combined cost by going to Bumper Bargains – you'd pay £4.77 + £2.45 = £7.22. (At Al's, you'd be paying £4.68 + £2.60, which works out at £7.28. At CutYourCosts Corner Mart, the total amount would be £4.86 + £2.40 = £7.26.)

p93: Working out interest

You should end up with two very similar amounts:

A €463.05: €400 × $(1.05)^3$ (ie $1.05 × 1.05 × 1.05$)

B €462.33: €380 × $(1.04)^5$

CHAPTER 5

pp110–11: The shared birthday conundrum

- The chance that a particular person *doesn't* share your birthday is very high: 364/365, or 0.99726. But the chance that a second person also doesn't share your birthday is $(364/365 × 364/365)$, or $(364/365)^2$: about 0.99453. You need to get all the way to $(364/365)^{253}$ before this number drops below 0.5, to around 0.499523. This proves, somewhat counter-intuitively, that there need to be 253 people in the room before there is more than an even chance that one of them *does* share your birthday.

- What are the chances of your birthday being on 1 July, like mine? This answer is much more straightforward. You have a 1/365 chance of sharing my birthday, since you could have been born on any day of the year with equal probability. (For the sake of clarity, we're ignoring the extra complications thrown up by leap years.)

CHAPTER 6
p120: Zero's odd behaviour
The weatherman means to say it's going to feel a lot colder than today, but mathematically 0×2 is still only 0.

pp124–5: Mathematically multiplying permutations
To crack the cipher **Brx duh d qxpehu zlcdug!** you have to substitute each letter for the letter four places back in the alphabet: so D becomes A, C becomes Z, B becomes Y, and so on. Doing this reveals the vital information: **You are a number wizard!**

The reason why the permutations for the Enigma machine work out at $5 \times 4 \times 3 \times 26 \times 26 \times 26$ is as follows. Briefly, there are five possible choices of wheel for the first slot, and for each of these options there are 4 possibilities remaining for the second, and 3 for the third. This means that there are 60 different ways ($5 \times 4 \times 3$) of putting the wheels into the machine. But for each of these wheels, there are 26 possible positions (due to the 26 letters on each wheel), giving us the calculation above – over one million possibilities.

INDEX

FURTHER READING

This selection includes books that provide a more in-depth exploration of mathematics, as well as those that will help you to extend your mental capacities.

Acheson, David *1089 and All That*, Oxford University Press 2002

Bracey, Ron *I2 Power-Up: 101 Ways to Sharpen Your Mind*, Duncan Baird Publishers 2008

Bridger, Darren and David Lewis *Think Smart Act Smart: 101 Ways to Be Effective and Decisive*, Duncan Baird Publishers 2008

Eastaway, Rob *Out Of the Box: 101 Ideas for Thinking Creatively*, Duncan Baird Publishers 2007

Gardner, Martin *Mathematical Magic Show*, Penguin Books 1986

Mankiewicz, Richard *The Story of Mathematics*, Weidenfeld & Nicolson 2000/ Princeton University Press 2004

Moore, Gareth *Keep Your Brain Fit: 101 Ways to Tone Your Mind*, Duncan Baird Publishers 2009

Pappas, Theoni *The Joy of Mathematics*, Wide World Publishing (US) 1993

Pickover, Clifford A. *A Passion for Mathematics*, John Wiley & Sons 2005

du Sautoy, Marcus *The Music of the Primes*, Harper Perennial 2004

Tipper, Michael *Memory Power-Up: 101 Ways to Instant Recall*, Duncan Baird Publishers 2007

AUTHOR'S WEBSITE

Andrew Jeffrey spends his life convincing people that they can do more mathematics than they believe. He also works for independent corporations who wish to communicate their corporate message at trade fairs and conferences in a fun and engaging way. For further details, visit Andrew's website at **www.andrewjeffrey.co.uk**

AUTHOR'S ACKNOWLEDGMENTS

I am hugely blessed to know quite a number of extraordinary people. I am a great believer in surrounding myself with people who are all better than me in some way (or many ways) and all of the following are examples of that philosophy!

My family and friends who have put up with me being largely absent over the past year – this book is the result of my occasional neglect, and I hope it was worth it.

Dr Tony Wing, my first mathematics tutor, who by never answering any of my questions as a student forced me to think through pedagogical and mathematical issues in a way that nobody else has ever quite managed since.

Caroline Ball, a brilliant editor, and one of the very few people capable of bringing any sort of order to my chaotic thoughts. More importantly than all that, she is a wise, kind and extraordinary person. The talented Katie John has also added her insight and direction – three heads have been infinitely better than one.

Rob Eastaway, a friend and fellow maths evangelist, whose suggestion it was that I do this in the first place! We share a love of mathematics and I am always encouraged by the time we spend together.

Stephen Froggatt, fine friend, a fellow magician, mathematician, musician and nutcase, with whom I have shared many beers, card tricks, curries and fascinating bits of algebra, sometimes simultaneously!

And finally to my wife Alison, who is more beautiful than she knows, and without whom it would all frankly be a little bit pointless.